VOLUME I
THE CONTENT CORE

SCOPE, SEQUENCE, AND COORDINATION OF SECONDARY SCHOOL SCIENCE

VOLUME I
THE CONTENT CORE

Revised Edition

A PROJECT OF
THE NATIONAL SCIENCE TEACHERS ASSOCIATION
1993

Copyright © 1993 by The National Science Teachers Association

The National Science Teachers Association
Scope, Sequence, and Coordination of Secondary School Science
1742 Connecticut Ave., NW
Washington, DC 20009

This material is based upon work supported by the National Science Foundation under Grant No. TPE-9050053.

Any opinions, findings, conclusions, or recommendations expressed in this material are those of the authors and do not necessarily reflect the views of the National Science Foundation.

Library of Congress Card Catalog Number: 92-060225
ISBN: 0-87355-107-9
NSTA Stock Number: PB-92

Edited by Marcia K. Pearsall
Revised book design by Marcia K. Pearsall
Cover photo © The Image Bank, Ilene Astrahan

Revised Edition
First printing
Printed in the United States of America

AN INVITATION

This document, the revised edition of **The Scope, Sequence, and Coordination Content Core**, was created to guide designers of new science curricula. The document also was developed with an awareness of individual teachers considering modifications in their courses or teaching methods.

The Content Core is intended to be a flexible document capable of encompassing and responding to varied educational needs and situations. Numerous scientists, science educators, science supervisors, and teachers participated in its development. But, even with this revised edition, this document necessarily is imperfect. It awaits the tests of flexibility and utility for those designers and teachers restructuring their science programs and courses across the nation. Can this document incorporate and support imaginative curricula under development? Does it provide a suitable framework for existing creative instructional materials? Does it aid in the development of new curricula? As designers and teachers begin to answer these questions with **The Content Core** in hand, the document will continue to evolve, undergo change, and be revised.

To these ends, then, the editors of **The Content Core** still extend an invitation for any person(s) to comment, critique, and propose changes to this document. We ask that you share both your successful and unsuccessful expressions of scope, sequence, and coordination so that, ultimately, others involved in restructuring can benefit. Our goal is for further editions of **The Content Core** to reflect what educators in the field discover of its strengths and weaknesses. Please direct your communications to:

Scope, Sequence, and Coordination
The Content Core
National Science Teachers Association
1742 Connecticut Avenue, NW
Washington, DC 20009-1171

FOREWORD

The Scope, Sequence, and Coordination Content Core serves as a guide for the design and construction of science curricula. **The Content Core** is not itself a curriculum, but rather, it organizes the subject matter of science according to the tenets of scope, sequence, and coordination described herein. Educators and teachers can use this document as a template for designing courses, selecting instructional materials, and constructing assessment instruments.

The Content Core is divided into two sections. The first section presents the principles of scope, sequence, and coordination and describes how they relate to science education restructuring efforts. The "Introduction" specifically identifies troubling aspects of science education in the United States and how SS&C addresses those problems. Individuals interested in further researching science education and the impetus for science reform are directed to the bibliography at the end of this document and to current education periodicals.

"Coordination of Science Content" discusses how designers and teachers might restructure science courses and programs using SS&C. This chapter proposes methods for integrating and coordinating the science disciplines. The list of organizational options is not exhaustive, and certainly designers are encouraged to develop means of integration and coordination appropriate to the unique needs of their students. However, SS&C does prescribe that schools offer the four science disciplines every year for seven years and believes that how schools relate and connect topics from each discipline is critical. In subsequent editions of **The Content Core**, this chapter will incorporate particularly imaginative and successful examples of integration and coordination to assist those engaged in the restructuring process.

Two examples of sequenced and spiralled topics follow the general coordination chapter. In the first essay, Arnold Strassenburg presents a sequenced model for teaching the concept of energy with particular attention given to moving from concrete experiences to abstract concepts. In the second essay, Russell Aiuto proposes a sequenced treatment of the phenomenon of inheritance. Aiuto's piece places greater emphasis on coordination with other disciplines. Finally, in a third essay, Bill G. Aldridge discusses the basic components of the natural sciences and their appropriate sequence in science programs.

The experiences and suggestions of the six SS&C pilot centers—located in California, Iowa, Puerto Rico, North Carolina, Texas, and Alaska—provide the basis for "Strategies for Implementation." This chapter also relies on reports from school districts across the country that are attempting various degrees of restructuring. "Strategies for Implementation" identifies some common elements of successful science programs and some frequently encountered obstacles to reform.

The second major section of **The Content Core** presents the proposed SS&C science content for each discipline: biology, chemistry, Earth/space science, and physics. Topical tables introduce each discipline and organize content into three grade level groups: 6-8, 9-10, and 11-12. Narratives follow each discipline table or set of tables. The narratives correspond to and elaborate each entry in the tables. Most narratives describe approaches for presenting content—descriptive, empirical, quantitative, and theoretical—and provide specific activity ideas. The narratives also emphasize important issues, principles, or concepts that underlie each topic. Ultimately, these narratives should stimulate teachers or course designers to innovate methods to engage students in the suggested science topics.

The Content Core contains a significant amount of science content, but was designed to maximize flexibility and to allow for different curriculum options. Each column of the discipline tables contains subtopics that can be distributed over a few years. For example, topics in each table's first column can be distributed over three years (Grades 6-8). Also, content within a column can be shuffled, particularly to achieve coordination among disciplines. For example, in Grades 9-10, "human interaction with the environment" in biology could be scheduled to fall in the same time frame as "human use of resources" in Earth/space science.

The sequencing of topics according to grade level is perhaps the most important aspect of **The Content Core**. The SS&C discipline teams deliberately sequenced the science content so that descriptive and phenomenological approaches begin the study of science in the middle level grades. If teachers find alterations in the sequence to be necessary, they should postpone topics in Grades 6-8 to later years, not "drop down" topics from Grades 9-10 or 11-12 into Grades 6-8.

Also, **The Content Core** can assist teachers and curriculum designers in identifying instructional materials consistent with an SS&C program. As change occurs in science education, textbook publishers will redesign their offerings, pilot sites will develop instructional units, and experts will create data bases describing educational materials. **The Content Core** provides criteria to evaluate the suitability of these new instructional materials.

The Content Core can help identify appropriate student assessment instruments. An SS&C curriculum, based on constructivism and inquiry-based learning, does not lend itself to traditional multiple-choice testing. As experts create performance-based assessment instruments, **The Content Core** offers a template to identify those which effectively assess student understanding of science concepts.

Finally, following **The Content Core** does not require schools to supplant curricula in development or frameworks in use. **The Content Core** serves as a guide to those individuals committed to restructuring their educational "delivery system" and has the flexibility to be hybridized with current curricula. The narratives contain activity suggestions but primarily aim to stimulate creativity. **The Content Core** is particularly compatible with the direction, tenets, and themes of the American Association for the Advancement of Science's Project 2061. **The Content Core** quite consciously reflects the 2061 theme that "less is more." This project concurs that covering science material using an encyclopedic approach does not promote the depth of understanding that is the hallmark of science literacy. Subsequent editions of **The Content Core** will correlate the mathematics standards developed by the National Council for Teachers of Mathematics.

FOREWORD TO THE REVISED EDITION

It has been a little over a year since the first edition of **The Content Core** appeared. Since the book's initial release at the National Science Teachers Association National Convention in Boston in late March 1992, the SS&C project has distributed almost ten thousand copies. The general response has been favorable, and it is clear that **The Content Core** has been of significant importance to a number of school districts, many of which were just beginning to restructure their science programs.

However, the project was aware that **The Core** was an imperfect document, and that, had not our SS&C pilot centers needed the book to go ahead with new course designs, we would have published a more complete guide for curriculum designers. In fact, many who have examined **The Content Core** have said that "it is a good first effort," or "we wish you could have expanded the discussions of how it could be used," or "some sections need more work." We have noted all of these observations and have found them to be reasonable criticisms.

It is in that spirit—the desire to improve what seems to be a useful guide—that we undertake this modest, early revision.

It is important to mention that since March 1992, the project has published a second volume in the SS&C series: **Relevant Research** (November 1992). This is a collection of papers that support the intellectual underpinnings of the SS&C reform, and it is, in effect, an expanded bibliography from Volume I reprinted under one cover.

Accounts of how **The Content Core** has been used in science programs thus far would be useful to new curriculum designers (i.e., examples of curricula designed with its guidelines, how teachers' time was scheduled, or how it has been used with other teaching materials). But, the project has decided to limit the detail on implementation to that which directly relates to restructuring science curricula with **The Core**. We plan to publish a third volume, **Implementing Reform**, in the latter half of 1993. This third volume will be a collection of case studies that embrace all elements of "how to" restructure a science program using SS&C, including curricula.

All material from the first edition has been reviewed and, where appropriate, revised or augmented. The first section has been expanded to include three articles that address how science content can be viewed, organized, and presented.

CONTENTS

INTRODUCTION

"Now, what I want is, Facts. Teach these boys and girls nothing but Facts. Facts alone are wanted in life. Plant nothing else, and root out everything else. You can only form the minds of reasoning animals upon Facts: nothing else will ever be of any service to them. This is the principle on which I bring up my own children, and this is the principle on which I bring up these children. Stick to Facts, sir!"

Charles Dickens, **Hard Times**

For generations, teachers have subjected students to educational programs designed in the tradition of Dickens' schoolmaster. By administering batteries of multiple-choice tests, teachers prodded students to achieve education's primary goal: the mastery of facts. Schools, teachers, and parents, educated in this tradition themselves, felt compelled to perpetuate this mind-numbing experience and to construct an educational edifice of facts.

Schools in the United States still practice and subscribe to this tradition. Educational policy makers, textbook publishers, and educators all reinforce the fact-driven educational structure. State legislatures mandate that students achieve certain educational objectives as measured by standardized tests. When schools and teachers are evaluated according to their students' performance on these multiple-choice tests, learning frequently becomes equated with successful memorization of test material. The emphasis on factual recall profoundly affects how teachers teach and, perhaps more significantly, how and what students learn. Accountability, to which state mandates are directed, seems a valuable component of the American educational system. And certainly, designing an educational program that emphasizes learning outcomes is admirable and desirable. But state legislatures need to take a hard look at how their policies ultimately affect student learning.

Modern textbooks reflect a preoccupation with facts as well. Even publishers that claim to offer progressive science textbooks—"breaking down the discipline boundaries"—retain an overwhelming amount of factual information and simply alternate chapters among the disciplines. Science textbook publishers admittedly follow the policy of "inclusion" because, they argue, "this may be the students' last chance to learn this material."

And finally, some modern educators still champion the 'mastery of facts' doctrine. E.D. Hirsch, for example, argues in his book, *Cultural Literacy*, that the decline in American education results from a neglect of factual information. Hirsch asserts that if students learned more shared facts, they would create a much-needed cultural context for themselves.

The typical U.S. science program discourages real learning not only in its overemphasis on facts, but in its very structure which inhibits students from making important connections between facts. Most science programs in U.S. secondary schools are organized in what is commonly called "a layer cake." Students study biology in the ninth or tenth grade, then chemistry the following year, and finish with physics in the twelfth grade. In a single year, students pursue one discipline from the descriptive to the theoretical, with little reference to prior science experiences—either in that course or other science courses—and even less reference to upcoming science experiences. One consequence is that many students never participate in those future science experiences: Three-fourths of American high-school graduates do not take science after the tenth grade (or, in layer-cake terms, after biology). The emphasis on facts and rote learning and the difficulties students encounter in grasping theoretical considerations without a grounding in experience deters many from continuing in science.

Clearly, the time has come for the educational edifice of facts and the layer cake to be dismantled. The United States is poised, both in terms of resolve and resources, to undertake science education restructuring. Teachers across the nation have indicated a readiness to participate in science education restructuring. Researchers already are experimenting successfully with new learning methods that do not require students to memorize a virtual encyclopedia of facts, terms, and bits of unconnected information. For years, the scientific community has asked the nation to reflect on the need for scientists and scientifically-literate citizens in our technology-driven world. And a consensus is developing among science educators that "less is more," i.e., that by covering fewer topics, students can develop a deeper understanding of science. If students learn a few concepts in depth, then they can apply them to new situations or problems. Students also must see how science directly relates to their lives and larger human concerns.

In order to achieve such comprehensive educational ends, this project advocates presenting key science concepts, appropriately sequenced, manageable in their scope, and coordinated within and between the science disciplines. This project, initiated by the National Science Teachers Association, is called **Scope, Sequence, and Coordination of Secondary School Science (SS&C).** The SS&C project specifically targets middle level students to generate an interest in science during their formative years and to encourage them to continue studying science.

SCOPE: A coherent science curriculum should span all six or seven secondary school years and involve all students. Curriculum designers should be guided by the "less is more" principle.

SEQUENCE: Science programs should involve appropriate sequencing of instruction, taking into account how students learn. Students should encounter concepts, principles, and laws of science at successively higher levels of abstraction over several years, making it possible for them to learn and understand science. Students need to experience the natural world before they learn the terms, symbols, and equations that scientists use to explain it.

Science programs should utilize spaced learning, such that teachers cover fundamental science concepts over years, not weeks or days. Repeated experiences in different contexts assist students in building concepts.

An SS&C curriculum also incorporates practical applications of science—engaging students in the early grades with science problems and issues of personal concern, and in the later grades with more global science considerations. Gradually putting science in a larger context helps students relate science to themselves and their lives.

COORDINATION: Biology, chemistry, Earth/space science, and physics share topics and processes. Coordination among these four disciplines leads students to an awareness of the interdependence of the sciences and their place in the larger body of human knowledge.

Specific methods for coordination are suggested in the following chapter.

Additionally, SS&C advocates that instructional strategies be appropriate for heterogeneous groups, with no tracking. SS&C believes that a diverse classroom improves learning. Students on all ability levels can exchange ideas and learn from each other. Since experiences precede the mastery of terminology, a constructivist approach to learning that responds to student preconceptions is most appropriate. Student-centered lessons, with emphasis on hands-on activities, are integral to the SS&C approach. Science programs should help students to answer the questions of science, not by presenting assertions or authority-determined answers, but by allowing them to propose and pursue the ideas, concepts, and information. Teachers should encourage students to ask:

- How do we know?
- Why do we believe?
- What does it mean?

The Content Core alone will not motivate students to ask these questions, to struggle with difficult science concepts, and to think clearly and logically. How curriculum designers and teachers use **The Content Core** and how educators assess the learning that occurs will affect student learning outcomes. As with present curricula, how and when the study of science involves quantification is critical.

Finally, designing and implementing an integrated and coordinated science program is a complex task. Advocates for restructuring must talk with state legislatures and educators. As publishers reassess the needs of the market, curriculum designers and teachers may need to take the initiative to compile and/or develop appropriate SS&C materials. SS&C encourages educators to press for materials that meet their needs and believes that such negotiations advance the goals described above. Scientists, educators, teachers, parents, and community members must join together to support science education restructuring efforts. The final goal is ambitious, but not unrealistic: Science learning for all students that is interesting, relevant, challenging, and personally rewarding.

COORDINATION OF SCIENCE CONTENT

The coordination component of SS&C involves interrelating the major ideas of biology, chemistry, Earth/space science, and physics.

The Content Core accommodates different coordination patterns, but resulting curricula should present science as first and foremost a way of (1) learning about the behavior of the universe and the matter and energy it contains, (2) organizing this knowledge so that it is comprehensible and useful to humans, and (3) developing models and theories about this behavior that not only correlate with past observations but also help predict future events.

Integrated Courses

Coordination can be achieved by designing and teaching a single, integrated course in science. Designers would establish a course focus to identify appropriate topics and the principles that could connect those topics. Three proposed single, integrated courses are outlined below. Again, **The Content Core** accommodates other designs; these are offered simply as illustrations.

Great Ideas of Science

A course organized around a Great Idea (or Ideas) of Science necessarily integrates those disciplines that operate under its laws or principles. The notion of "Great Ideas" suggests that scientists arrived at a single, encompassing idea after making extensive observations and working through complex chains of thought. Therefore, course designers must construct Great Idea courses carefully to connect the experiences of children with the larger concept. Teachers should not start with the Great Idea and work backward. Research indicates that students learn science better if they move through concrete experiences to abstractions. The challenge is to design activities that both teachers and students find intellectually and educationally manageable.

Evolution: Evolution could guide a Great Idea course. When studying biological aspects of evolution, students could explore variation among organisms, schemes for classifying them, and their growth and reproduction. Activities would address habitats and their relationship to species survival and evidence, such as the fossil record, that confirms and clarifies evolutionary processes. This course would examine natural selection critically as the dominant mechanism driving evolution.

5

In class discussion, students might talk about how theories become accepted as dogma until or unless current or more consistent evidence challenges their viability.

The evolution course also would incorporate Earth/space science. Activities could provide evidence for the Earth's internal structure and age. Students might consider the relatively recent discovery that ocean floors are constantly replenished at some locations and returned to the Earth's interior at others. They might speculate how the existence and behavior of volcanoes contribute to those processes. The course would introduce the theory of plate tectonics, the evidence supporting the idea of continental drift, and how earthquakes are caused by this process.

Finally, students' knowledge of evolutionary processes on Earth would allow them to conceive of these processes at work in the universe. Students would contemplate the Milky Way as a clump of stars called a galaxy and then the enormous number of existing galaxies so far away that the light they emit takes billions of years to reach Earth. Scientists who examine the light from these galaxies conclude that they are moving away from Earth, but that these galaxies originally shared the same location. Such observations suggest to scientists that the universe started with a Big Bang. Astrophysicists, in particular, have worked out the details of what probably occurred microseconds and millennia after the Big Bang. Astrophysicists also use their data to predict what will happen to our universe in the future. From this physics component of the evolution course, students could learn about the birth and death of individual stars—expanding and building on their knowledge of gravity, electromagnetic interactions between light and matter, nuclear fusion, and the elements in stellar and interstellar matter.

Energy: Like evolution, energy could guide a Great Idea course. Physicists, chemists, Earth scientists, and biologists all rely on its principles. Energy principles are critical to the analysis of particle kinematics and dynamics and to the study of electromagnetic theory, thermodynamics, atomic and nuclear physics, and quantum mechanics. Chemists describing chemical interactions, the rates of reactions, the stability of molecules, and the production and decay of transuranic elements must understand energy.

An energy-focused course could address biological topics such as photosynthesis, respiration, and fermentation as these processes involve the sources and sinks of energy in cells. On a larger scale, students should conceptualize growth and reproduction as energy transfer processes on both the molecular and the cellular levels.

Earth scientists analyze energy issues when they study erosion, weather phenomena, ocean currents and tides, and plate tectonics. For example, to reconcile the Earth's age and known heat transfer properties of matter with the continued existence of thermal energy within the Earth, scientists recognized radioactivity as the major source of that energy.

Astronomers and astrophysicists constantly raise questions about the energy sources that fuel phenomena such as the supernova and radiation emissions from quasars. To understand the energy source for sunlight itself, scientists had to

6

recognize nuclear fusion as a process consistent with the conditions within a star and the rates at which stars emit electromagnetic radiation.

In the next section of **The Content Core**, Arnold Strassenburg offers a model of how this energy concept could be sequenced and spiralled through the middle and high school years.

Phenomena

A course organized around phenomena allows for the integration of disciplines that contribute to the understanding and occurrence of the phenomena.

Space Exploration: Space exploration serves as an umbrella concept for introducing topics from diverse fields of science and—if one wishes—social science. Course designers might set up introductory activities that explore space and time: What exists out there between the planets and stars? How far apart are the objects humans might want to visit? How long would such a trip take?

The course could present the mechanics of space travel. Students might explore mechanisms that could drive a space ship for long periods of time, energy sources available to the space ship, and issues of astronauts' health and safety. What effect might weightlessness, a restricted diet, limited physical activity, and increased levels of radiation have on bodily processes such as respiration, blood circulation, and digestion? Can humans tolerate stresses such as large accelerations on take-off and landing, close quarters with others, and limited mental stimulation for long periods of time?

Lastly, students would try to state the declared, and perhaps undeclared, goals of the space enterprise. They would need to consider the role science and technology plays in setting and achieving those goals. What do humans wish to learn? What kinds of instruments do they need to collect, store, and transmit data? How should society compare and assess the risks and benefits of space travel? Students might conclude that society should scrap the idea of human passengers and begin again to design robotic explorers.

The Production, Distribution, and Consumption of Food: A course on the production, distribution, and consumption of food would incorporate issues and principles familiar to physicists, agricultural scientists, chemists, and biologists. The course would raise questions such as: What is a food? How is it produced? What does it do for the organisms that consume it? How do the plants that humans consume extract minerals and other substances they need from the soil and the atmosphere? Students could consider how these substances are replenished, and what happens when they are not. They might try to assess whether humans improve on nature by selective breeding.

Students should observe what happens when foods are cooked and what chemical changes take place when foods are digested. Students could conduct experiments to determine the "food energy" calories required for the mechanical energy to walk up stairs or to run a marathon. Then students could identify the processes by which muscles and bones use "food energy."

With some imagination, designers could structure several courses around a central phenomenon or phenomena. The course designers' task is not to generate topics to study, but to organize and integrate existing science topics such that the whole is greater than the sum of the parts. The phenomena need not be all-encompassing. Courses could even include a series of individual phenomena that, taken together, make sense. The Houston SS&C pilot center produced a series of "blocks" for the seventh and eighth grades that studied phenomena such as floating and sinking, heat and temperature, and adapting to the environment.

On pages 17-24 of **The Content Core**, Russell Aiuto provides an example of how phenomena could focus a course or serve as a theme throughout a science program. He proposes a sequenced treatment of the phenomenon of inheritance. He includes some suggestions for coordination with other disciplines as well.

Science, Technology, and Society

A Science, Technology, and Society course integrates disciplines under the rubric of science in its modern context. The challenge for those designing STS courses is not finding appropriate issues, but rather selecting a set of issues that encompasses all the science concepts included in **The Content Core**. The Iowa SS&C pilot center used an STS pattern of integration to create "modules" such as "Mr. Dream House (energy)," "You are What You Eat," and "Flight."

Automobile travel: Automobile travel could focus a single, integrated STS course. Students initially might analyze design features and their relationship to auto travel. Auto-users demand certain performance standards in terms of acceleration, fuel economy, and quality of exhaust. Auto-users also must balance their desires for speed, comfort, safety, and style.

The course could examine elementary kinematics and dynamics. Activities would raise questions such as: What determines the maximum acceleration? At this acceleration, how long does the auto take to reach a particular speed? How do engineers determine that a particular curve is safe at 30 mph? Students could consider various ways to propel a car. Since gasoline pollutes and its long-term supply is uncertain, students might question why electric cars are not in common use. Could engineers design a solar-powered car or a car that burns hydrogen? What determines the horsepower rating of a conventional car? Because students tend to be concerned about the environment, the course should include a unit on automobile exhaust. Students would identify pollutants in car exhaust and suggest why they are hazardous to human health. Students could compare methods for minimizing pollution.

Automobile use relies on many auxiliary structures and services. Activities might center on how engineers could design roads to optimize automobile travel or how automobile travel in congested areas could be made tolerable. Mass transit might be proposed as a better solution, but students then would need to assess mass transit options.

Environmental quality: Like automobile travel, environmental quality could organize an STS course. In different regions of the country, the topics could vary.

In coastal regions, the preservation of wetlands and ocean pollution might be major topics. Students would study the biological organisms that occupy wetlands and the ocean. They would learn about the chemical and physical properties of ocean water and how organisms have adapted to these conditions. Students would examine the natural processes that produced and continue to modify the landforms and sur-rounding oceans and the effect that human activities have had on these natural processes.

Certain STS courses would require students to handle more complex science knowledge and experiences and sort out the competing claims of science and society. For instance, school districts that lie east of heavily industrialized regions might design a single, integrated course around acid rain. Students would study the chemistry of smokestack gases and acid-base reactions. They could examine the water cycle and the properties and motions of the atmosphere. Students could learn about trees and how they produce food and grow or organisms that live in lakes and why changes in water pH threaten the organisms' existence. Most significantly, students would confront the perplexing issues that can arise when scientific evidence comes in conflict with diverse human interests.

Similar to acid rain, global warming could act as a course focus, but its conflicting data would need to be as carefully presented as its scientific premises. Students would look at the composition of the atmosphere, its "original state," and how it is changing. The role of deforestation in the carbon cycle might guide an activity. Students should consider what physical and chemical processes cause the Earth's temperature to increase when atmospheric carbon dioxide increases. In the presence of increased carbon dioxide and warmer temperatures, shouldn't plant life flourish? Students would need to examine ozone and its "disappearance" from the atmosphere. They could assess the threats to human health this disappearance poses.

Discipline-Based Courses

Science programs can achieve coordination not only through integrated courses, but also through disciplined-based courses. Designers, though, must structure these discipline-based courses with an awareness of the criticisms of the traditional layer-cake science curricula. SS&C proposes two discipline-based models consistent with SS&C principles.

Discipline-Based Courses Taught in Parallel

The "pure" version of this model entails four separate courses—biology, chemistry, Earth/space science, and physics—taught simultaneously by teachers best qualified in each discipline. Students would attend one or two periods each week in each discipline. (The periods might change from year to year, i.e., more Earth science in Grades 6-8, more physics in 11-12). Coordination results from frequent confer-ences among the teachers. For example, the physics course might explore kinetic theory at the same time that the chemistry course introduces ideal gas laws. The Earth science course examines atmospheric pressure's role in determining weather patterns, and the biology course investigates osmotic pressure as a way of pumping sodium ions through cell membranes.

Similar to the kinetic theory focus in the above example, the four disciplines could coordinate activities around light. The physics teacher introduces the concept of light propagation as a photon stream. The chemistry teacher discusses photochemical reactions, and the biology teacher presents photosynthesis. The Earth/space science teacher then could provide evidence for the Big Bang theory and note that the Doppler effect for light provides the critical evidence for Hubbell's law.

The logistics of scheduling and coordinating four separate, but simultaneous science courses places great demands on both the curriculum designers and teachers. A "less pure" coordinated science program would put two teachers in daily contact to teach simultaneously a physical science course and a biological science course. This model might work especially well in the lower grades.

Discipline-Based Courses Taught in Series

The series model achieves coordination by offering each discipline in 1/4-year segments such that students learn the same level of science material each year, encounter smaller "chunks" of knowledge, and draw connections more readily between disciplines. Thus, in one year, an individual student takes one quarter of physics, one quarter of chemistry, one quarter of biology, and one quarter of Earth/space science. The series model is perhaps most effective for districts that can offer only a few courses and teachers who feel comfortable teaching only in a specific discipline. The series model effectively reduces the thickness of the layers of the traditional layer cake curriculum. This series model, though, avoids certain problems of the layer cake and remains consistent with SS&C principles.

First of all, students enrolled in a series model science program study topics from each science discipline each year. Second, students explore a particular science discipline over several years. Distributing science content allows the student to benefit from the spacing effect and allows the teacher to treat a specific topic at different times over the years, each time increasing the level of sophistication. Finally, even if some students drop out of science before completing the entire sequence (if the school does not require a science enrollment every year or the student drops out of school before completing Grade 12), they still will have studied introductory concepts from the four disciplines. In 1990, the majority of students who completed high school in the United States did not study all four disciplines.

Summary

Curriculum designers and teachers can achieve coordination by using the above models, modifying those models, or designing programs suited to their students' particular needs. But, in addition to careful design and planning, an effective, coordinated science program should establish procedures for on-going review of the program and should involve dedicated educators throughout the first few years of implementation.

REVISITING THE ENERGY CONCEPT IN A SCOPE, SEQUENCE, AND COORDINATION CURRICULUM

Arnold Strassenburg

Energy is an important concept in physics, and it has applications in every discipline of science, every branch of engineering, and every field of medicine. For young students, introductions to energy should involve concrete examples of different kinds of energy possessed by macroscopic objects and systems. Eventually, in later grades, students should be able to identify several different kinds of energy, recognize processes that convert one kind to another, and know what it means to say energy is conserved. In addition, high school students should be introduced to subtleties such as these: (1) The thermal energy of a macroscopic object can be associated with the kinetic energies due to the random motions of the constituent particles. (2) The rest mass energies of nuclei can be associated with the kinetic and potential energies of individual nucleons. (3) The energies carried by and distributed throughout fields of electromagnetic waves are the same energies attributed, in a quantized form, to the photons in the particle model of electromagnetic radiation. (4) In thermodynamics, the primary concern is with how energy passes through the boundaries of a system, and heat transfer and work are recognized as the two fundamental mechanisms. Between the early experiences with energy and the later discussions of the abstract concepts listed in the previous sentences, students must progress through several levels of understanding. An SS&C curriculum should be designed to revisit the concept repeatedly over students' school years, each time helping them to achieve a higher level of understanding. A sketch is presented below to suggest how this upward spiraling treatment of the energy concept might be designed.

Early Experiences

Students should be given opportunities to have experiences that will support intuitive feelings about the concept of energy. It would help, in the beginning, to establish a simple criterion for when an object possesses energy. For example, one might set up a small battery on its circular base and declare that any object that can knock it down has energy.

Kinetic Energy

In the first activity, students could roll a cart toward the battery. On collision, the battery falls down. So, by the established criterion, the cart has energy. Clearly it is

11

the cart's motion that causes the battery to fall, so this kind of energy is named kinetic energy.

Further play with different carts moving at different speeds reveals that low mass carts with very low velocities fail to knock the battery over. However, a low mass cart with a large velocity does knock it over, and so does a cart with large mass moving at a very low velocity. So this activity should demonstrate to students that kinetic energy must increase with both mass and velocity.

Potential Energy

In another activity, students can hold a cart at rest on a ramp so that when the cart is released, it will roll down onto a tabletop, strike the battery, and knock it down. Clearly, the cart has energy. The new idea here is to suggest that the cart had that energy before it was released. As the cart rolled down, it gained kinetic energy. But the students need to see that, because the cart was elevated above the tabletop with the force of gravity acting on it, it had the potential to acquire kinetic energy with human intervention. This kind of energy is called "potential energy," or more precisely, "gravitational potential energy."

Activities should explore other kinds of potential energy. Compressed springs and bar magnets oriented perpendicular to the Earth's magnetic field provide convenient opportunities to show that a kind of potential energy can be associated with several different force fields.

Thermal Energy

There are many ways to demonstrate that hot objects have energy, and that the amount of energy increases with the temperature. One simple activity is directing a stream of steam coming from boiling water onto the battery so that the battery falls. The students probably will note that it is the motion (kinetic energy) of the steam that knocks the battery over. Teachers should help students see that the steam particles would not have acquired enough kinetic energy to knock the battery over unless the water temperature increased, so the system's energy did increase with its temperature.

Intermediate Experiences

The early energy concepts should be broadened in the intermediate years in three ways: (1) additional forms of energy should be identified; (2) mechanisms that transform energy from one form to another should be studied systematically; (3) quantitative measurements should be made that begin to suggest that when a certain amount of energy in one form disappears, an equal amount of energy in one or more other forms is generated.

Additional Forms of Energy

At minimum, students should be provided with opportunities to become familiar with these additional forms of energy:

Electrical energy: A seat of emf, such as a battery, produces an electric current in a circuit. If the current passes through a resistor, the resistor gets hot. Since activities already have established that hot objects have a form of energy called thermal, this observation establishes that the battery-driven current has energy, and this kind of energy is called electrical.

Chemical energy: Many chemical processes produce readily noticeable amounts of heat. Once again, it can be inferred that the chemicals possessed the energy before they interacted to produce the heat. This kind of energy is called chemical energy.

The energy of electromagnetic waves: Sunlight, when absorbed by solar collectors, heats water in the collectors. When focused by a converging lens, sunlight can ignite bits of paper. So, once again, it can be concluded that sunlight carries energy.

It is important to convince students that all light, not just sunlight, carries energy. For example, a beam from a light bulb can be used to open and close doors. A bit harder, but still important, is to introduce the idea that nonvisible forms of electromagnetic radiation carry the same kind of energy. At one end of the electromagnetic spectrum, radio waves can produce electric currents in radio receivers that are eventually converted to sound. Nearer the other end of the spectrum, x–rays pass through human bodies and cause chemical changes in photographic film. At this point, for students to accept the idea that all these forms of electromagnetic radiation are related is largely an act of faith. Eventually, when the different forms' respective behaviors are linked in the theoretical structure called the electromagnetic theory, students will have a better understanding. But the electromagnetic theory should not be introduced until the later years of high school.

Energy Changes From One Form To Another

Throughout the investigations focusing on energy, the idea that energy changes from one form to another should be emphasized. This is such an important idea that efforts should be made to explore many examples in a systematic way. Here are a few suggestions:

• A swinging pendulum illustrates transitions from potential to kinetic to potential energy.

• The expansion of an object as the gas within it is heated illustrates thermal energy changing to either kinetic or potential energy or both.

• The heat generated in the brakes of a car as it is brought to rest illustrates kinetic energy changing to thermal energy.

• Sliding down a rope at constant speed illustrates the conversion of potential energy to thermal energy.

Formulas

To determine quantitative changes in the amount of energy an object or a system possesses requires that students know formulas for how energy depends on various measurable parameters. It would be inappropriate in the intermediate years to derive these formulas from theory, and inappropriate at any stage to simply write down formulas without justification. A sensible approach might be as follows:

• Students measure the temperature increase that results from heating a quantity of water in a container (that prevents heat loss to the surroundings) as a function of time. The graph of the data should be a straight line, showing that the temperature changes vary linearly with the amount of energy added to the system. Thus temperature change can be used as a measure of thermal energy.

• When lead shot confined to a long cardboard tube is turned end-for-end many times, the temperature of the shot increases. Measurements will reveal that the increase in temperature is proportional to the total distance the shot has dropped. This activity establishes that gravitational potential energy increases linearly with elevation above some reference level.

• Next students could roll carts with good bearings down a ramp and measure the speed at the bottom as a function of the drop in elevation. The data should strongly suggest that kinetic energy varies with the square of the speed.

• The next step might involve measuring the maximum speed of a mass connected to a spring oscillating horizontally as a function of the amplitude of oscillation. This measurement should establish that the elastic potential energy of a spring varies as the square of its extension beyond its normal length.

• Next, students could measure the maximum compression of a spring that is compressed by a cart of mass m moving at speed v. By varying both v and m, activities could reconfirm that kinetic energy increases as v^2 and establish that it increases linearly with m.

• Once students know a few such measures of common energy forms, teachers should schedule new experiments to check on the conservation of energy concept. Since complete formulas are still unknown, it is not suggested that students measure energy changes in joules. Instead, it is suggested that activities demonstrate, for example, that when water is heated by an electric current through a resistor two different times (keeping the time of heating and everything about the resistor and the water the same except the current through the resistor), the temperature change is larger by a factor of four in the second run if the current was doubled. Since students know from earlier experiments that the energy delivered by a current to a given resistor in a given time varies with the current squared, the new experiment is consistent with the idea that all the electrical energy supplied by the current appears as thermal energy in the water.

The Later Years

In the final two or three years of high school, teachers should direct students' attention to the overarching theoretical ideas that tie together many of the concepts introduced in earlier years and open up new ways of understanding phenomena that are inconsistent with classical theoretical structures such as Newton's laws.

Thermodynamics

The first law of thermodynamics helps students to clarify their ideas about energy in two ways: First, it reinforces the concept of energy conservation. Secondly, it directs attention to the mechanisms by which a system exchanges energy with its surroundings.

The concept of "internal energy" has a major role to play in thermodynamics. By emphasizing that this quantity depends only on the state of the system (its temperature and pressure, for example) and not on how it reached that state, thermodynamics invites students to make connections between macroscopic descriptions of the system energy (all forms lumped together as internal energy) and the energies possessed by the individual particles of the system (which are kinetic, potential, chemical, etc.).

Finally, the second law of thermodynamics helps students to make connections between energy exchange processes and make generalizations from the study of thermal physics such as "heat always flows from matter at a higher temperature to matter at a lower temperature." It also provides a theoretical framework for discussing practical applications such as improving the efficiency of heat engines and designing systems that minimize thermal pollution.

Nuclear Processes

It is strongly recommended that high school seniors be introduced to two nuclear processes: fission and fusion. Studying fission and fusion is valuable because it reveals to students that mass is a form of energy—an idea that does not emerge readily from the study of objects as large as atoms. Also, most students have intense curiosity about atomic and hydrogen bombs and about nuclear reactors.

From a theoretical viewpoint, mass as a form of energy emerges from the postulates of special relativity. Because special relativity is not recommended for inclusion in the high school science curriculum, the interchangeability of rest mass and other forms of energy must be introduced as a hypothesis that summarizes many observations of such changes as nuclear interactions. To enable students to understand why mass changes to other forms of energy, both when heavy nuclei fission and when light nuclei fuse, teachers must introduce the concept of nuclear stability. Then, it requires only a small conceptual leap to examine the cosmological evolution of elements in the stars. How far a class can go in this direction should be a local decision.

Quantum Physics

It is not appropriate for high school students to be introduced to the Schrodinger equation or taught the subtleties of wave mechanics. It is appropriate, however, that they read about and observe phenomena that reveal the particle–like properties of electromagnetic radiation and the wave–like properties of beams of small particles. A beam of photons, then, becomes another model for how radiant energy is transmitted. Previously, the students were introduced to the idea that electromagnetic waves carry radiant energy.

With this study, students complete their growing inventory of the major forms of energy and also glimpse the quantum world. In this quantum world, a particular result of an event, like nuclear decay, is probable, not certain. This statistical analysis of the way nature behaves will come as a revelation to students who previously had been exposed only to the doctrine of cause and effect relationships. It should impress students with the way scientific theory adapts to the observation of new phenomena and should prepare them for college–level courses.

SEQUENCING AND SPIRALLING THE SCIENCE CONTENT OF INHERITANCE

Russell Aiuto

This essay presents a possible sequence of twelve topics for study of the phenomenon known as inheritance. This sequence follows the historical development of the science of genetics. It is designed so that topics that are introduced early in the sequence are revisited later in new contexts. This sequence is intended for seven years of study. Topics 1 and 2 should be studied only in the middle level years and Topics 10 to 12 only in Grades 11-12.

The explanation for each topic is necessarily brief, but it is followed by a short discussion (in italics) which correlates the topic to the other science disciplines. The connections to chemistry and physics are fairly obvious for Topics 10 and 12, but it may be less apparent that Earth science principles can be correlated with Topics 1 to 3, physics with Topics 7 to 9, and mathematics throughout. Even more remarkable connections can be made between a specific topic and the other science disciplines if more detail is provided for the experimental activities that support the twelve "issues."

Topic 1: Continuous Versus Discontinuous Variation

In determining what is passed on from parent to child, and how it is transmitted, students need to look carefully at a characteristic. The historical and first inclination is to look at the entire organism at once, but a listing of characteristics or "impressions" alone will result in confusion. The second idea would be to identify a "trait" and to see how it varies from individual to individual. At some point, one will have to decide that a trait "is" or it "is not" (it is discontinuous), or that a trait cannot be separated into discrete classes given measuring instruments with infinitely small gradations (it is continuous). Measuring height of students and graphing the measurements or trying to separate a large sample of different oak leaves into "classes" would be useful experiences. In the case of the oak leaves, students will find both continuous and discontinuous (into different species) variation. It would not be too early for instruction to introduce simple statistical manipulations such as mean and standard deviation as well as a variety of pictorial representations of measurements such as graphs and bar graphs.

Discussion or activities should show that continuous and discontinuous variation are not unique to living systems.

17

The most important outcome of this study of continuous versus incontinuous variation is to demonstrate that chaos results from either trying to derive rules from quantitative traits or looking at the whole organism at once. The products of independent probabilities will show the infinite number of classes and the resulting confusion. The reason that empirical laws of genetic transmission could be formulated was because Mendel and some of his predecessors concentrated on a few discontinuous traits: tall/dwarf and red/white.

The concept of continuous and discontinuous variation is an important one in all science disciplines. How humans describe objects and phenomena is crucial to their understanding of them. In a sense, this early science instruction is getting at issues of "qualitative" versus "quantitative." The rudiments of measurement, variation around a mean, and dimensionality can be applied to various science topics including size and shape, temperature, and states of matter.

Topic 2: Genetically-Determined Versus Environmentally-Influenced Variation (Lamarckianism)

Activities on this topic should involve consideration of the following questions: How do we know that a trait is caused by heredity? Why could it not be caused by nutrition, exercise, or moisture? And if variation can be caused by the environment, why cannot that change be transmitted to offspring? Uniform garden experiments with plant clones will help students begin to address these questions. Students probably also need to test current cultural myths about the effect of environment on heredity. Students will need time to work out well-established misconceptions. Growth experiments that take several months are not inappropriate.

Some wonderful exercises in logic are related to consideration of genetically-determined versus environmentally-influenced variation. Why are mutilations not inherited if the environment can "change" heredity? Discussion could consider Weissman's experiment with cutting off twenty generations of mouse tails. Students could be challenged to think about what difference exists between the genetically-determined tailless Manx cat and dogs that have their tails docked.

An interesting extension to the above explorations is to examine the problem of high cholesterol. Is it caused by diet? Heredity? Both? How would you find out if going on a low cholesterol diet will decrease blood levels of cholesterol? For some people, the diet will decrease cholesterol. For others, it will have little effect. This example can be revisited when the biochemistry of gene action is studied.

Clear connections can be drawn from this topic to photosynthesis and all of its physical properties: soils and soil types, mineral nutrition, and development.

Topic 3: Discrete Characteristics Versus Blending

Students need to examine discrete characteristics versus blending. Are there rules that can be derived when looking at such traits as hair color or flower color? How do you know that some traits that are intermediate shades between two extremes are not the result of the "mixing" of genetic determinants? What evidence is there to support the idea that whatever the genetic determinants are, they do not lose their individuality in subsequent transmissions? Using colored pieces of glass or plastic, such as theatrical filters, it is possible to construct alternative models for blending versus discrete units. Also, pedigree analysis will reveal that some characteristics skip generations rather than produce intermediate individuals with an "intermediate trait." A particularly interesting challenge is the markings on the leaflets of white clover (*Trifolium repens*) which can vary both in the extent of the stripe, its position, and, in some cases, its color. Is it possible to predict what the parents might have looked like using a blending hypothesis? Could there be two different traits, one "blending" (the extent of the stripe) and one retaining a "discreteness" (the position of the stripe)? The clover system provides excellent material for population genetics as noted below in the coordination with the study of evolution.

Since the particulate nature of genetic material is an essential point in these examinations, the "integrity" of basic units, or their inability to be subdivided except by unusual intervention, is an idea that can be found in all of the science disciplines. In a way, the concept of reductionism and its validity can be examined across the disciplines.

Topic 4: Following A Trait

There are a number of human traits that can be followed over at least three genera-tions: free and attached ear lobes; eye color; hair color; hair texture; pattern baldness, with a prediction for the student. (Some caution should be used in following ear lobes as they are not the best trait, and following eye color usually requires oversimplification and arbitrary grouping.) There are several illuminating points for students to consider and/or learn: (a) a trait may not always be present in a single generation; (b) a trait may or may not be associated with gender; (c) if a trait appears in one of the later generations, where could it have come from?

In a very broad sense, the concept of continuity is being introduced in these explorations and, with the idea of constancy introduced above, can provide raw material for the study of evolution, the consideration of stability versus instabil-ity, and issues in other disciplines.

Topic 5: Deriving Rules From Chance Events (Mendel's Laws of Segregation and Independent Assortment)

Students never derive Mendelian laws. They are assertions given to them in illustrations in textbooks—usually Punnett squares jazzed up in some way.

The problem with the usual treatment of these "laws," which really are model extrapolations from quantitative data, is that they are pictured as precisely four gametes from a meiosis, or, worse yet, a single series of results, precisely following a famous Mendelian ratio. What needs to be understood is that the kinds of gametes are, on the average, of a certain kind, and that offspring, on the average, fall into statistical categories known as Mendelian ratios. Students should begin to develop the understanding that "batches" of gametes and offspring are involved and that approximations of probability are involved.

The importance of "pure lines," i.e., that crossing experiments must start with homozygous parents, needs to be derived by students. They should not simply be told, "Well, you have to start with pure lines."

As students begin to derive the Mendelian rules, teachers should be aware that this may be, unlike Mendel, their first experience in expanding the binomial.

Another difficulty is that students have to accept the Mendelian outcomes on faith, in that they generally are given just the F2 generation—green:albino corn seedlings growing comfortably in flats. The true test would be for students to grow up the green seedlings, self-cross them, and determine the frequency of green corn plants that produce only green and the frequency of green corn plants that produce both green and albino seedlings. That, of course, is impossible. There is no way out of this experimental dilemma unless students are allowed to make crosses and can follow a trait through the F2 and a testcross. Fruit flies can be used for this purpose, such as wild-type versus vermilion eye, but in recent years there has been a disinclination to bother with the trouble of maintaining cultures. Perhaps the crosses could be made for them (or before them) and the harvest of offspring left for the students.

Some schools may have the facilities to maintain garden plots of soybean mutant strains or long-term colonies of *Tenebrio* variants or some other system. The most promising system, and one that appears to be expanding as additional strains are developed, are "Fast Plants" with a sufficient number of mutant types now available to permit genuine discovery by students.

Chance and chance events are important components of many science phenomena. Diffusion, osmosis, and, in a very important sense, concepts such as gas laws and temperature-related phenomena have probability relationships. Students can begin to consider chance versus design and the relationship of empirically-derived laws to probability-related generalizations.

Topic 6: Explaining Apparent Exceptions To Rules

Given that students might obtain results that do not conform to the Mendelian rules of simple statistical ratios—for example, ratios that depart from 3:1 and 9:3:3:1, how can activities demonstrate that these exceptions are merely variations of the basic rules? Epistatic ratios and linkage are the two simple, but interesting, exceptions. It is not the deviation from the rules of independent assortment that is important here. Rather, it is (a) the interaction of gene pairs to produce epistatic ratios (which will be returned to when gene action is studied), and (b) the spatial relationship of genes to the linear configuration known as the chromosome. Epistatic maize ears are challenging and can permit students to formulate explanations for these, at first approximation, "crazy" numbers.

Historically, linear recombination maps were made independently of chromosome maps. Students are going to be faced with the same dilemma. Intuitively, they know that there is a correspondence of the two, but how do they prove it? That is part of the challenge in studying topics 7, 8, and 9.

This topic has great potential for examining some fundamental issues in science. For example, scientists often find what appear to be exceptions to the rules or laws or extensions or elaborations of highly predictable phenomena. When that happens, though, they might obtain additional evidence from an entirely different viewpoint. Secondly, there are indeed examples in the history of science of exceptions that have given rise to entirely new avenues of investigation, e.g., Bateson's admonition to "treasure your exceptions." Going even further, this topic can generate the examination of the idea of paradigm shifts, e. g., in astronomy's shift from a Ptolomean to a Copernican system.

Topic 7: Correlating Statistical Events With Cellular Events (Sutton-Boveri Hypothesis)

Now, students can begin model building in earnest. If students understand mitosis and meiosis, at least in their fundamental behaviors, it should be possible for them to make the next intellectual leap. The segregation of chromosomes at Anaphase I of meiosis and Anaphase II of meiosis explain independent assortment. This can be illustrated with different colored "chromosomes" and then with chromosomes marked with genes. Even though material to perform experiments cannot be found, it should be possible to propose them. Pipe cleaners should not be used; different colored rubber tubing with connectors is preferable because later they can be used for model building of crossover products.

The ultimate evidence for this subject would be first and second division segregation of spores in *Neurospora* or *Sordaria* . These spore activities also would contribute to the depth of understanding of Mendelian laws since examining the results of a haploid system without the diploid rations of 3:1 forces some serious analysis.

These activities raise questions about statistical correlations in all sciences: When do statistical correlations become sufficiently compelling so as to constitute evidence? What reliability can be placed on "cause and effect" phenomena? Earth/space science provides an example: the correlation between continent shape and continent origin is, in itself, an intellectual inference requiring evidence.

Topic 8: Qualitative Versus Quantitative Significance Of Chromosomes

If chromosomes and chromosome action can be correlated with the Mendelian units, and if chromosome number is important in normal development, then the next question for students to consider is: Is there a "bulk" requirement of chromosomes—a certain number—but without qualitative distinction? Or, does a creature have to have not only a certain number of chromosomes but a certain number of different kinds of chromosomes? How could one distinguish between these two possibilities?

Differences in chromosome number in humans, and the resulting syndromes, are interesting phenomena to students, but unless students understand how chromosomes are gained or excluded in gamete production, such syndromes become exercises in clinical descriptions. Non-disjunction, as exemplified by Bridges work (1916) with fruit flies can be hooked to the next topic: X-linked inheritance as the ultimate correlation of genes (as followable traits) with chromosomes. It is entirely

21

possible to consider topics 8 and 9 as a single item, but it may be more productive to think of both of them as spiralling topic 7.

Alternative hypotheses, of which this is one, abound in science. How does one obtain evidence to distinguish between two superficially plausible explanations? Acceleration and speed are examples in physics, i.e., they allow students to confront an "easy" explanation versus an experimentally-determined one. Phenomena for which common metaphors are constructed, such as phototropism, can provide experiences in resolving naive versus rational explanations.

Topic 9: Demonstrating The Cellular Basis Of Heredity (X-Linked Inheritance)

Pedigree analysis of human X-linked traits such as defective color vision, hemophilia, and glucose-6-phosphate dehydrogenase deficiency would serve to illustrate the cellular basis of heredity. (White-eyed fruit flies also would provide an excellent illustration.) One could extend this topic to more complex contexts later, e. g., X-linked genes in humans have interesting biochemical actions.

The mechanism of sex determination often is of interest to students. Some challenging scientific questions include: (a) Does the Y chromosome determine maleness? (b) Does the X chromosome, in absence of the Y, determine femaleness? (c) Are sex determination systems in other animals the same?

Although verification has been a theme throughout the study of science, this topic is particularly suited to raising the question, How do we know? The everyday, commonplace nature of the questions asked, "What determines what I am?," can be found in all science disciplines and can be examined at different levels of complexity. The explanations, as in all of science, require verification.

Topic 10: Redefining The Gene (one gene = one enzyme)

Thus far, students have roughly defined the gene as a followable trait in a breeding experiment. They should now consider, what does a gene do? In other words, having defined it in terms of transmission, how do we define it in terms of function? Two lines of evidence can be used in activities: (a) nutritional mutant strains in yeasts or molds; (b) study of biochemical pathways with accumulation products, as occur in PKU and alkaptonuria in humans.

Historically, (b) preceded (a) in that the work of Garrod (just after the rediscovery of Mendel in 1900) proposed that a number of "inborn errors of metabolism" were caused by errors in Mendelian genes. Pedigree analysis and analysis of laboratory findings from samples of afflicted individuals are activities that can be undertaken. Then, the more recent "one gene = one enzyme" work of Beadle and Tatum can be demonstrated with appropriate microorganisms. Please note, though, that it is more important for students to arrive at a working model of how a gene functions, and how that model could be tested, than to repeat the Beadle and Tatum experiments.

Direct correlation to chemistry and physics is possible with this topic: the nature of biochemical pathways, oxidation-reduction reactions, the physics of enzyme action, and the ubiquity of the gene-enzyme-metabolic function system.

Topic 11: Identifying The Gene

Activities for identifying the gene should involve more than just building a model of the double helix. Students need to think about, of all the various chemical compounds within a cell, which is "the genetic molecule?" What are the requirements for such a molecule? Since students have correlated genes and chromosomes, they know it has to be one of the substances that make up chromosomes. What are they? Once the substance has been identified, how can its structure support the requisites of genetic function?

The identification of the genetic material, experimentally, began with Avery in 1944, with experiments that fragmented bacteria into their component molecules and then attempted to determine which of the molecules had the biological property indicating that it was the genetic molecule. The study of these experiments alone (even if no laboratory activities can be identified for this purpose) is extremely valuable, since there are critical steps in the Avery experimentation that raise interesting questions. Meselson and Stahl's experiments with bacteriophage infection are important to examine in this exploration as well.

Examination of the work of Chargaff needs to precede the construction of the Watson-Crick double helix.

Students should look at how the structure of DNA demonstrates its ability to replicate itself as well as provides information for the production of the gene product.

There are a number of interesting books about "the race for the double helix" for those teachers and students who enjoy reading about the personalities of science. Comparing Watson's **The Double Helix** with Sayre's **Rosalind Franklin and DNA** is fascinating sociology of science, particularly when viewed against the dispassionate treatment of "the race for the double helix" by Judson, **The Eighth Day**, or Olney, **Genetics and DNA**.

Obvious connections exist between this topic and chemistry and physics, but some important opportunities for examining bonding forces, polarity, encoding, and x-ray diffraction also exist. However, more importantly, these activities can illustrate the importance of model-building in discovery since the achievement of Watson and Crick really is their use of the experimental data of others to build the double helix. Corresponding examples can be found in the other science disciplines.

Topic 12: Manipulating The Gene

Now, and only now, should students (equipment and resources permitting) engage in recombinant DNA studies.

An important scientific issue here is that gene splicing is the fulfillment of a theoretical prediction. The history of this biotechnology is basically the search for the mechanisms, steps, and techniques that would be clearly necessary for the fulfillment of the model. It is so systematic in its approach and development that it suggests that the direction of science is the realization of predictions. What other examples are there in chemistry, Earth/space science, and physics?

A second and important point that can be made is that "biotechnology," as represented by recombinant DNA studies, is a striking expression of the integration of the science disciplines of biology, chemistry, and physics. Another expression is the technology attendant to space science.

BASIC COMPONENTS OF THE NATURAL SCIENCES

Bill G. Aldridge

The basic components of the natural sciences include the processes used, and the resulting products of these processes, that are common to all natural sciences. Those common processes are: observing, classifying, measuring, interpreting data, inferring, communicating, hypothesizing, developing models and theories, and predicting. The products of these processes are terms, facts, concepts, principles, empirical laws, theories, and applications, as well as the instruments and tools of science.

Unlike processes used in the natural sciences, the resulting products often are identified with particular subject fields or disciplines. In order to be considered a basic component of all natural sciences, a product of science must transcend a particular subject area or discipline. It must be a fundamental component in each major subject area: physics, chemistry, biology, and Earth/space science.

Basic Terms Common To All Natural Sciences

Terms are: names of entities, objects, specific events, specific time periods, classification categories, organisms, or parts of organisms. Because of their specificity, terms are almost always unique to a particular science subject field or subfield. There are very few terms that are basic components of all natural sciences. Terms are used mainly as tools for communication and generally are learned as needed for reading or communicating science. Those terms that are common to all natural sciences include:

Multiples associated with the prefixes nano, micro, milli, centi, kilo, mega, giga

Names of units of time, mass, length, area, and volume

Names of SI derived units for force, electric charge, pressure, magnetic field, work, energy, and power.

Basic Facts Common To All Natural Sciences

There are basic facts common to all natural sciences. These facts are usually operational definitions, measurements, or observations—all of which can be replicated. Such facts should not be memorized intentionally, although through repeated use, a person will remember many such facts. Instead of memorizing these facts, a person should understand the underlying concepts for each fact and where to look up the fact in a reference. Some important facts that are common to all natural sciences, and the grade levels where they are first to be encountered, can be specified. It is understood that the level of abstraction or the degree to which the fact is made quantitative increases with grade level, so that such fundamental facts are revisited at those higher levels in a spiral fashion. The list is as follows:

Grades 6-8

Definition of an inch in centimeters

The speed of light in a vacuum

Value of absolute zero on the Celsius temperature scale

The density of liquid water

Boiling point of water (2 significant figures)

Freezing point of water (2 significant figures)

Speed of sound in air (2 significant figures)

Speed of sound in water (2 significant figures)

Definitions of minute, day, hour, and year

Names of common acids and bases

Names of common elements that are electrical insulators

Names of common elements that conduct electricity

Grades 9-10

Wavelengths in nanometers of visible colors of light (2 significant figures)

Approximate range of wavelengths in nanometers of ultraviolet and infrared radiation (2 significant figures)

Refractive index for glass and for water (2 significant figures)

Components of simple, diatomic molecules of common gases and common compounds, in terms of numbers of atoms and molecular mass

Numerical value of 1 mole

Volume of 1 mole of gas at STP

Definition of 1 atomic mass unit

Atomic numbers and atomic masses in grams per mole of a few common elements: oxygen, nitrogen, carbon, and hydrogen

Approximate dimensions of inorganic molecules

Components of common atoms and isotopes in terms of protons and neutrons in nucleus and electrons surrounding nucleus

Value of calorie in joules

Heat capacity of water

Latent head of vaporization of water

Latent head of fusion of water

Value of acceleration due to gravity on Earth (2 significant figures)

Definition of a pound in newtons

Atmospheric pressure on Earth (2 significant figures)

Approximate radii of the Earth, moon, and sun

Grades 11-12

Definition of an electron-volt in joules

Definitions of simple, face-centered, and body-centered cubic crystal structures

Permittivity and permeability of free space

Half-life of carbon-14

The charges on common free element ions, in units of charges on the electron

Approximate radii of an atom and of an atomic nucleus

Value of Planck's constant

Boltzmann's constant

The constant, k, in Coulomb's law of force for electric charges

Mass of, and charge on, the electron, proton, and neutron

Value of the constant, G, in the law of universal gravitation

Value of mass-energy equivalence

Value of the ideal gas constant, R

Basic Concepts Common To All Natural Sciences

For the purposes of this paper, a science concept is defined as follows:

A regularly occurring natural phenomenon, property, or characteristic of matter which is observable or detectable in many different contexts, and which is represented by a word or words and often by mathematical symbols. Most science concepts are derived from others (i.e., speed derived from concepts of distance and time). When a derived science concept is in the form of an equation, it is a mathematical definition, not a natural relationship (e.g., mass density).

There are many basic concepts of the natural sciences that transcend any one subject field or discipline.

When placed in a list, these basic concepts appear to strongly emphasize physics, and to a lesser degree, chemistry. That is because physics is such a basic science, and one upon which much of chemistry is built. Chemistry and physics are fundamental to an understanding of the life sciences. And chemistry, physics, and the life sciences are fundamental to an understanding of the Earth and space sciences.

The basic concepts that are common to all natural sciences, and the level at which they are first encountered or used, can be specified. These basic concepts are encountered at almost every level of science education. However, in the earlier years, they are considered in qualitative, descriptive ways. As the school level increases, each concept is revisited at a greater level of abstraction, and the contexts of their use become more diverse. They also become more quantitative. That list is as follows:

Grades K-5

distance

depth

length

width

height

perimeter

circumference

angle

solid

weight

liquid

conservation or being conserved

Grades 6-8

scalar (a number and a unit)

precision (and significant figures)

accuracy (and calibration against a standard)

exponent

power of ten and order of magnitude

continuity and discontinuity

length, area, volume, mass, electric charge, and time

surface area

mass density

time rate of change with distance (speed)

time rate of change with speed (acceleration)

time rates of change in general

velocity (speed and direction)

force

pressure

buoyancy

work

energy

power

kinetic energy (translational and rotational)

potential energy (gravitational, electrical, and nuclear)

pulse

wave (monochromatic)

wavelength

amplitude

superposition of waves

diffraction

interference

spectrum

reflection

refraction

dispersion

period

periodic motion

simple harmonic motion

equilibrium

thermal equilibrium

temperature

heat

thermal energy

freezing point

boiling point

conductivity

concentration

mixture

homogeneous

element

compound

solvent and solute

acid

base

crystal

gas

Grades 9-10

vector (a number, a unit, and a direction)

power

atom

molar volume

molecule

atomic number

mole

atomic mass

nucleus

heat capacity

isothermal

catalysis

colloid

current (mass, charge or volume per unit of time)

distance rates of change (contours and gradients)

evolution

centripetal

gravity

invariance

longitudinal wave

momentum (linear and angular)

osmosis

pH

radioactivity

resonance

electric field

magnetic field

transverse wave

electromagnetic wave

monochromatic

intensity

Grades 11-12

displacement (distance and direction)

intensity

vector acceleration

wave velocity

polarization of waves

coherence (temporal and spatial)

ion

charge density

torque

electric dipole

adiabatic

internal energy

enthalpy

entropy

general concept of density as distribution function

latent heat of fusion

latent heat of vaporization

chemical potential energy (as a special case of electrical)

elastic potential energy (as a special case of electrical)

fission

fusion

half-life

isotope

photon

plasma

simultaneity

relativity

Basic Empirical Laws Common To All Natural Sciences

Similar to the basic concepts of the natural sciences, there are basic empirical laws common to all natural sciences. For the purposes of this paper, an empirical law is defined as follows:

A generalization of a relationship that has been established between two or more concepts through observation or measurement, but which relies on no theory or model for its expression or understanding (e.g., pressure of a gas as a function of volume, holding temperature and number of moles constant).

Many empirical laws were once the subject of theories and, indeed, may still be of such concern. It is the observed phenomena, behavior, or process that characterizes the empirical aspect of the law. Explaining the law is the proper subject of theories. For example, during an interaction of two bodies, the product, *ma* , for one body has the same magnitude as the product, *ma* , of the other. This is an empirical law. Explaining that law is the task of a theory. The following empirical laws are common to all natural sciences. Levels shown are those at which the empirical laws are first encountered or used. These laws may be used at several levels of school science, but will become successively more quantitative in their use at the higher grades.

Grades 6-8

Pascal's law

Boyle's law

Charles' law

Gay-Lussac law

Law of specular reflection

Law of conservation of energy

Evolution (observed patterns of change over time)

Grades 9-10

Law of limiting factors

Photosynthesis (as a process)

Graham's law

Chemical periodicity (as observed, not from theory)

Laws of definite and multiple proportions

Newton's three laws of motion

Observed variation of strength with linear dimension change

Inverse square law

Ampere's law

Coulomb's law

Faraday's law

Gauss's law

Snell's law of refraction

The Doppler effect

Observed variation of mass with linear dimension changes

Observed variation of surface area with linear dimension changes

Grades 11-12

Le Chatelier's principle

Hess's law

Dalton's law of partial pressure

Law of universal gravitation

Law of conservation of momentum

Observed variation of heat through a surface with linear dimension change

Wien Displacement law

Fourth Power law of Radiation

Photoelectric effect

Theories And Models Common To All Natural Sciences

Theories are used to explain facts, observations, phenomena, and empirical laws. Theories often incorporate various concepts that have been quantified and that use symbols in their representation. Thus such theories are often mathematical. To best understand a theory, one needs to experience, either directly or vicariously, the process by which earlier theories are created, hypotheses derived from those theories tested, and the evolution of such theories over time. It is important to recognize the fact that theories are tentative, and even the most recent, are merely in one stage of a continuous process. It is also important to recognize that theories, unlike facts, observations, and empirical laws (that summarize data, measurements, or observations), are creative constructs which do not necessarily map one-to-one with reality. Many alternative theories might explain the same set of phenomena. In that sense, the theories are not a reflection of reality and may well be considered subjective.

There are those in science education who do not make the important distinction between facts, observations, and empirical laws as one class of knowledge, and the theories and models scientists create to account for those facts, observations, and empirical laws as another class of knowledge. The former are expressions of objective reality. The latter are admittedly subjective creations of the mind, and to the degree that such theories make the same predictions, they are equally valid. The best theories are the ones that make the most comprehensive testable predictions. But empirical laws and facts drawn from replication of independent observations of natural phenomena are in no way subjective. Examples like the discovery of helium first on the sun, through spectral evidence, and the subsequent use of spectroscopy to analyze distant stars and galaxies relies strongly on the universality and congruence of empirical law and nature.

The fact that theories make predictions and that those predictions often result in a new empirical law merely corroborates the theory. The theory never is proved. But the resulting empirical law stands until such time that replication fails to produce the same result or results.

Theories, more than any other expression in science, can be described at many different levels of abstraction. Also the sequence of hypotheses and theories leading to recent theories is very important to emphasize in science education. Students can learn such theories at a very descriptive level, later at a somewhat more symbolic level, and so forth, until they are using very sophisticated mathematics. The following theories, and the level at which they are first encountered or used, are common to all natural sciences.

Grades 6-8

Wave theory of light (When light behaves wavelike)

Grades 9-10

Theory of scaling laws (surface area, volume, and mass)

Theory of the chemical bond (from early theories of atoms and molecules resulting from laws of definite and multiple proportions to theories associated with various kinds of chemical bonds)

Atomic theory (from sequence of Thomson, Rutherford, and Bohr models)

Grades 11-12

Electromagnetic theory (as synthesis of empirical laws of Ampere, Faraday, and Gauss, with new prediction on speed of waves, with connection to light and confirmation by Hertz)

Theory of heat (Invention of concept of internal energy, first law as semi-empirical, second law as theory, following use of postulates and mathematical reasoning)

Theory of scaling laws (periodicity, heat transfer through surfaces and strength)

Kinetic-Molecular theory (as explanation of relationship of temperature to average kinetic energy of molecules)

Particle (photon) theory of light (when light behaves particle-like)

Quantum theory (atomic and nuclear, including orbitals, quantum numbers)

The modern Periodic Table (from theory, including Pauli exclusion principle)

The modern theory of solids (include elements of solid state physics as it applies to crystals, metals, and semi-conductors)

Cosmology and theories of evolution (explanations of general evolutionary change)

Applications Of Basic Science Concepts, Empirical Laws, And Theories

In a very real sense, the life and Earth sciences constitute natural applications of fundamental concepts, principles, and laws of science, most of which appear in physics and chemistry. These natural applications are in contrast to the applications of basic science to create human-made devices, like those in the arts, music, manufacturing, transportation, or consumer products for the home.

Studies of the design, creation, and use of technology is normally the subject matter of engineering or technology and not of natural science. There is, of course, considerable overlap between science and technology, especially as technology forms the instruments and tools of science. As one learns science, there are many instruments and tools which depend upon basic concepts, principles, and laws of the natural sciences.

Facts, concepts, empirical laws, and theories from the various natural sciences may form important components of many personal and societal problems. Although such science should be studied, and time devoted to learning these applications, it is essential that time scheduled for learning the natural sciences not be used for decision-making exercises in social science, health, economics, political science, civics, philosophy, or religion. This does not preclude a careful and detailed examination of the science associated with serious personal or societal problems or issues. Nor does it preclude interdisciplinary studies when time and expertise is provided for teachers in science and these other fields to coordinate the learning experiences in their respective subjects. But these studies should not include subsequent social action or decision-making which is more properly civic responsi-bility, political or religious involvement, or the content of other school subjects. Social action and such decision-making exercises just should not be part of education in the natural sciences.

Given the fact that basic concepts can be approached initially from a qualitative, descriptive level and later from a quantitative, more abstract level, the sequencing problem is not at all formidable.

At the lower levels of abstraction, as in the middle grades, the same teacher can teach both the basic sciences and the applications in the life, Earth, and physical sciences. At the high school level, where science is learned more abstractly and quantitatively, specialists need to work as a coordinated team.

Since most of the life, Earth, and space sciences are, themselves, complex applications involving facts, concepts, laws, and theories drawn from basic laws of the natural sciences, a clear coordination of topics in those high school subjects must be made with the underlying natural sciences. Not only are the life, Earth, and space sciences largely applications of natural sciences, but so is much of what is learned in secondary school physics and chemistry. Such applications of basic laws to specialized subject areas will be considered below.

Applications of science also include the various tools and instruments of science. The following application sections assume that basic concepts, empirical laws, and theories are learned concurrently or as needed to approach the application.

Grades K-5

At the elementary school levels, mainly concepts are considered, with much experience provided in different contexts. Also, process skills are developed, with many opportunities for creating and testing hypotheses, offering descriptive explanations for observations (or for patterns of observations), and building primitive theories.

The major problem at this level is that teachers and textbooks often want to offer the modern, "correct" theories or explanations, rather than allow the explanations developed by students to stand when those students have reached the limit of their resources or abilities. If the criterion, "How do you know?" is applied carefully, this unfortunate practice can be curtailed.

In terms of subject matter not common to all natural sciences, K-5 students should acquire a repertoire of experience with the daily, monthly, and annual positions of the Earth, sun, and moon, with some ability to describe what they are likely to see in terms of the locations and appearances of these objects. They might also begin to see patterns in the stars and notice changes in those patterns over a year. Students also should have experience with simple substances and living organisms.

Concepts That Are Common To All Natural Sciences That Are First Encountered Or Used In Grades K-5:

distance

depth

length

width

height

perimeter

circumference

angle

solid

weight

liquid

conservation, or being conserved

Grades 6-8

At this level, young people can concentrate on additional experience with a wider range of science phenomena, develop numerous empirical laws from experimental evidence, and construct hypotheses and theories to explain the phenomena and laws they have found.

At the end of Grade 8, many of the process skills should be well-developed, and experimental procedure thoroughly understood.

As in K-5 levels, there is the unfortunate tendency of texts and teachers to emphasize modern explanations and theories and to discount the limited explanations or theories developed by students at this level. This problem must be corrected.

Concepts That Are Common To All Natural Sciences That Are First Encountered Or Used In Grades 6-8:

scalar (a number and a unit)

precision (and significant figures)

accuracy (and calibration against a standard)

exponents, powers of ten, and order of magnitude

continuity and discontinuity

length, area, volume, mass, electric charge, and time

surface area

mass density

time rate of change with distance (speed)

time rate of change with speed (acceleration)

time rates of change in general

velocity (speed and direction)

force

pressure

buoyancy

work

energy

power

kinetic energy (translational and rotational)

potential energy (gravitational, electrical, and nuclear)

pulse

wave (monochromatic)

wavelength

amplitude

superposition of waves

diffraction

interference

spectrum

reflection

refraction

dispersion

period

periodic motion

simple harmonic motion

equilibrium

thermal equilibrium

temperature

heat

thermal energy

freezing point

boiling point

conductivity

concentration

mixture

homogeneous

element

compound

solvent and solute

acid

base

crystal

gas

Empirical Laws Common To All Natural Sciences That Are First Used Or Encountered In Grades 6-8:

Pascal's law

Boyle's law

Charles' law

Gay-Lussac law

Law of specular reflection

Law of conservation of energy

Evolution (observed patterns of change over time)

Theories Common To All Natural Sciences That Are First Used Or Encountered In Grades 6-8:

Wave theory of light (when light behaves wavelike)

Grades 9-10, 11-12

Assuming as prerequisites the levels of understanding of the basic concepts, empirical laws, and theories that are common to all natural sciences and the experiences described previously, the high school experiences in natural science should focus on the areas listed below. These listings in no way prescribe how the learning is to occur. Indeed, there are numerous ways that the learning may occur. An examination of the scientific or technical components of personal decisions and societal issues should be part of the learning experience.

Ninth Grade Chemistry

solutions

enzymes/catalysts

types of reactions

equilibria I (vaporization, condensation, dissolution, and precipitation)

simple stoichiometry

Ninth Grade Biology

organismal biology

a) photosynthesis (observational evidence)

b) respiration (substrate utilization/ digestion)

Mendelian genetics I

a) principles

b) use of probability

adaptations (evidence of evolution)

response to environment (evidence of evolution)

the optical magnifier and microscope

Ninth Grade Earth/Space Science, Geology

soils

mapping

contours

fuels

Oceanography

ocean and weather

Astronomy

the refracting telescope

stars (patterns)

measurements by parallax

measurements by inverse square law of light

Meteorology

weather instruments

weather

air masses

climate

Ninth Grade Physics

rockets and trajectories

machines

simple calorimetry

thermal insulation

Concepts Common To All Natural Sciences First Used Or Encountered In Grades 9-10:

vector (a number, a unit, and a direction)

power

atom

molar volume

molecule

atomic number

mole

atomic mass

nucleus

heat capacity

isothermal

catalysis

colloid

current (mass, charge, or volume per unit of time)

distance rates of change (contours and gradients)

evolution

centripetal

gravity

invariance

longitudinal wave

momentum (linear and angular)

osmosis

pH

radioactivity

resonance

electric field

magnetic field

transverse wave

electromagnetic wave

monochromatic

intensity

Tenth Grade Chemistry

pH meter

types of chemical bonds

oxidation/reduction

gas reactions

molar calculations

common crystals

Tenth Grade Biology

centrifuge

simple chromatography

Leibig's Law Extended

cellular biology (cell structure and function)

genetics II

a) extensions

b) exceptions

chromosomes

a) mitosis

b) meiosis

c) Sutton-Boveri, Bridge

genetic versus environmental variation (evidence of evolution)

Tenth Grade Earth/Space Science, Geology

faults, stresses folding-deformations

earthquakes

uniformity

Oceanography

sonar devices

tides

marine habitats

ocean circulation

coriolis effects

Astronomy

planetary features and detection

Meteorology

condensation and precipitation

heat transfer

global circulation

Tenth Grade Physics

thermal expansion

sound waves and resonance

Doppler effect

single lens and reflector optics

simple electric circuits

Empirical Laws Common To All Natural Sciences First Used Or Encountered In Grades 9-10:

laws of limiting factors

photosynthesis (as a process)

Graham's law

chemical periodicity (as observed, not from theory)

laws of definite and multiple proportions

Newton's three laws of motion

observed variation of strength with linear dimension change

inverse square law

Ampere's law

Coulomb's law

Faraday's law

Gauss's law

Snell's law of refraction

The Doppler effect

observed variation of mass with linear dimension changes

observed variation of surface area with linear dimension changes

Theories Common To All Natural Sciences First Used Or Encountered In Grades 9-10:

theory of scaling laws (surface area, volume, and mass)

theory of the chemical bond (from early theories of atoms and molecules resulting from laws of definite and multiple proportions to theories associated with various kinds of chemical bonds)

atomic theory (from sequence of Thomson, Rutherford, and Bohr models)

Eleventh Grade Chemistry

spectrophotometer

electro-chemistry, half-cells, and redox

chemistry of carbon

natural polymers (carbohydrates, lipids, proteins)

Eleventh Grade Biology

phase contrast microscope

cellular biology

a) photosynthesis

b) respiration: glycolysis and fermentation

biochemical genetics

a) one gene = one enzyme

b) mutation (evidence of evolution)

populations

a) structure

b) dynamics, e.g., Hardy Weinberg, genetic drift

Eleventh Grade Earth/Space Science, Geology

landform

evolution (geological changes)

coastal processes

Oceanography

changing levels of the coastal region

glaciation

Astronomy

meteorites

asteroids

Meteorology

storms—hurricanes and tornadoes

optical phenomena of the atmosphere—rainbows, mirages, and halos

Eleventh Grade Physics

AC electric circuits

batteries

motors and generators

meters

optical instruments

prisms and diffraction gratings

Concepts Common To All Natural Sciences First Used Or Encountered In Grades 11-12:

displacement (distance and direction)

intensity

vector acceleration

wave velocity

polarization of waves

coherence (temporal and spatial)

ion

charge density

torque

electric dipole

adiabatic

internal energy

enthalpy

entropy

general concept of density as distribution function

latent heat of fusion

chemical potential energy (as a special case of electrical)

elastic potential energy (as a special case of electrical)

fission

fusion

half-life

isotope

photon

plasma

simultaneity

relativity

Twelfth Grade Chemistry

biochemistry (enzymes, biosynthesis)

equilibria II (reaction rates)

polymers, artificial

Twelfth Grade Biology

electron microscope

molecular biology

a) photosynthesis (carbon dioxide fixation, light dark reaction, Calvin-Benson cycles)

b) respiration (Kreb's cycle)

molecular genetics

a) structure of DNA

b) central dogma

c) recombinant DNA

theories of evolution

a) before Darwin

b) natural selection

c) mechanisms

d) speciation

e) extinction

Twelfth Grade Earth/Space Sciences, Geology

theory of plate tectonics

nuclear counting technology

radioactive dating

landform evolution

coastal processes

Earth interior

Bowen's reaction

series

Astronomy

tracking reflecting telescopes

spectrometer

radiotelescopes

formation of stars and planets (evolution theory)

planetary interiors (laws and theories)

Oceanography

ocean basin development

Meteorology

atmospheric phenomena

carbon dioxide and greenhouse gases

ozone creation and depletion

climate change

Twelfth Grade Physics

devices utilizing nuclear fission and fusion

metals and solid state materials

applications of electromagnetism

heat engines and refrigerators

Empirical Laws Common To All Natural Sciences First Used Or Encountered In Grades 11-12:

Le Chatelier's principle

Hess's law

Dalton's law of partial pressure

law of universal gravitation

law of conservation of momentum

observed variation of heat through a surface with linear dimension change

Wien displacement law

fourth power law of radiation

photoelectric effect

Theories Common To All Natural Sciences First Used Or Encountered In Grades 11-12:

theory of scaling laws (periodicity, heat

transfer through surfaces and strength)

kinetic-molecular theory (as explanation of relationship of temperature to average kinetic energy of molecules)

electromagnetic theory (as synthesis of empirical laws of Ampere, Faraday, and Gauss, with new prediction on speed of waves, with connection to light and confirmation by Hertz)

theory of heat (invention of concept of internal energy, first law as semi-empirical, second law as theory, following use of postulates and mathematical reasoning)

particle (photon) theory of light (when light behaves particle-like)

quantum theory (atomic and nuclear, including orbitals, quantum numbers)

the modern Periodic Table (from theory, including Pauli exclusion principle)

the modern theory of solids (including elements of solid state physics as it applies to crystals, metals, and semiconductors)

cosmology and theories of evolution (explanations of general evolutionary change)

STRATEGIES FOR IMPLEMENTATION OF SCOPE, SEQUENCE, AND COORDINATION

Conditions for Successful Restructuring Efforts

Restructuring science programs successfully requires change and accommodation on the part of both science teachers and school administrators. Issues of scheduling, teacher training, parental support, and program assessment pose particular challenges to those involved in the restructuring process. In particular, the SS&C recommendation that all students take science every year for seven years requires creative scheduling and difficult administrative decisions. Nonetheless, vital new science programs will emerge across the nation when individuals make such changes that initially appear undesirable and impossible to accomplish.

Inservice teacher training must occur concurrently with curricular change. For middle level teachers, adopting SS&C entails teaching across the four science disciplines, rather than focusing on the traditional life sciences or Earth sciences. This shift may require teachers to supplement their knowledge in less familiar subject areas. Interacting with colleagues who teach other science subjects will be critical to maintaining a properly sequenced and coordinated science program.

Thus, school districts serious about restructuring their science programs must commit to a well-designed and well-supported program of professional development for their science teachers. Summer workshops and one-day workshops offered regularly during the academic year are necessary components of this professional development program. Further, teacher inservice programs should be diagnostic: they should provide content experiences for teachers consistent with middle level curriculum materials, not replicate university-level science syllabi.

Since specific science subjects may be taught only a few periods a week, homework and out-of-school explorations act as essential extensions of the classroom. Homework assignments maintain continuity in student learning. To ensure community support for SS&C efforts, school personnel should meet with the parents of all students before SS&C classes begin. School personnel should explain the project's objectives and its benefits for their children. Parents should be informed of the different expectations for out-of-class work and be apprised of how they can help students meet these expectations.

An SS&C program should shift the focus of learning science to the student and away from the textbook, tests, and even the teacher. Textbooks, tests, and teachers

41

should exist to guide the student, not control, dominate, or make the student feel overwhelmed by overly-detailed, technical, or excessive science information. Ideally, the teacher provides the context for learning.

The focus of learning science shifts to students when they can challenge and develop their own theories, collect their own data, and present their own outcomes of investigations. In other words, when the teaching process becomes a guiding process, students can take responsibility for their own education.

Many teachers rightly complain that existing models of science teaching encourage the use of textbooks that cover too much material. Some teachers respond by not using texts. While these teachers can work without a supportive framework, others cannot. The SS&C project asserts a direct relationship between students taking responsibility for learning and teachers taking responsibility for teaching. The SS&C project supports activities and curricula that relieve the teacher's obligation to be a holder of knowledge—encyclopedias can do that—and that allow the teacher to manage the learning environment. Science education surely will benefit when a new generation of textbooks, reduced in scope, but coherent in sequence, are developed to contribute to that learning environment rather than dictate its character.

Strategic Planning in Restructuring Science Programs

The conditions for a successful restructuring project must be met, and school districts, as with any organization seeking productive change, must establish a rational planning process. Too often educational reform takes place in a piecemeal fashion, with varying degrees of adjustment to individual components of the educational program.

Since numerous parties have a stake in effecting science reform, the SS&C project encourages all such stake holders to involve themselves in the restructuring process. Certainly parents as well as district administrators should involve themselves at every level. Community persons who look to the schools not only to provide employees, but to contribute to community life should involve themselves in the planning process. While such collaboration appears to complicate the restructuring process, it actually reduces the possibility that these constituencies will feel threatened and resist the changes sought.

Finally, scientists are key stakeholders who should involve themselves at every stage of program planning and implementation. Although designing precollege science education programs places additional demands on scientists, the SS&C project believes that the expertise of practicing scientists is critical to the restructuring enterprise. Not only do scientists verify the accuracy of subject matter content, but they serve as the primary source of inservice instruction for teachers.

Creating even the initial conditions for change must involve accounting for the goals of all concerned. Involved parties need to reach a consensus on the following questions: What do we aim to accomplish with our educational program? What obstacles stand in the way of achieving our goals? What strategies are necessary to remove these obstacles? What commitments must we make to remove these obstacles? What do we need to implement our program, and how do we acquire these resources if we do not have them?

Assessment of Student Achievement

Restructured science programs should increase student achievement and understanding while involving more students in science. But students' progress cannot be assessed unless assessment instruments are consistent with what students are expected to learn. Schools participating in SS&C must develop an assessment component to their new program to determine if student participation and achievement are, in fact, improving. Current classroom assessment exercises, for the most part, involve recalling facts and require little reasoning, thinking, or relating laws and principles. Program designers and teachers should review such classroom testing and evaluation procedures. Standardized tests and many teacher-designed tests distribute student scores to show variation in student achievement. These tests force half of the students to fall below the median in accomplishment without assessing the learning that did occur. The SS&C project calls for the assessment of students' qualitative progress and demands educators' agreement that teaching is directed at understanding.

An inquiry-based science program requires performance-based assessment. These assessment instruments measure successively higher levels of student performance and raise the question, "Are these levels adequate for these students at their current age and experience?" Outcome statements are made operational by the kinds of problems and tasks students can perform at different times during their school experience.

Several states, including California, Connecticut, and New York, have incorporated performance-based assessment in their science programs. The National Science Teachers Association, as an extension of its SS&C project, is currently developing a Compact Disc-Interactive system to assist schools in administering assessment components to their programs.

BIOLOGY SEQUENCE GRADES 6-12

Sub Topics	Grades 6-8	Grades 9-10	Grades 11-12
PROPERTIES OF LIVING THINGS			
Structures Unique to Life	basic cell structure at the LM level patterns of structure at the organismal level	cell functions and how cells are adapted to perform them organizational hierarchy: cells, tissues, organs, systems, organisms	biochemistry of life processes: respiration, photosynthesis, and protein synthesis
Interaction with the Environment	response to stimuli: tropisms, taxes, and homeostasis	the cell	
Reproduction	life cycles patterns of reproduction: asexual and sexual	reproduction of cells: mitosis and cell division meiosis inheritance	molecular genetics
Composition	chemical composition: carbohydrates, fats, proteins, and key inorganic compounds		

BIOLOGY SEQUENCE GRADES 6-12

Sub Topics	Grades 6-8	Grades 9-10	Grades 11-12
THE LIVING ORGANISM			
Systems	human organism human systems	diversity among organisms classification	growth and development differentiation
Cycles	human life cycle		
THE BIOLOGICAL PLANET			
Components	components: niche, habitat, population, community, ecosystem, and biome		
Interactions	organisms and the physical environment	interrelation-ships between organisms effects of humans on the environment	evolution: theories on the origin of life natural selection changing populations
Patterns	patterns of matter (cycles)	patterns of energy flow	patterns of evolution

PROPERTIES OF LIVING THINGS

Properties common to living things as well as properties unique to living things serve as the first biology content organizer.

Structures Unique to Life

Basic cell structure at the light microscope level: Curricula for Grades 6-8 should address basic cell structure as observed at the light microscope level. Students should identify the cell wall, cell membrane, nucleus, cytoplasm, chloroplast, and vacuoles. Activities should allow students to distinguish and compare characteristics of both the basic plant cell and animal cell. It is important that students engage the functions associated with these cell structures on a mechanical, rather than a biochemical level: nucleus and the control of cell function, chloroplast, and photosynthesis. They will need to examine several examples to be able to generalize that the cell is the basic unit structure of life. Students should encounter the microscope as a tool for enhancing the resolution of observation.

Activities should raise questions about cells. Why are cells so small? Using principles of volume, teachers should lead students to an answer. Activities can involve models—sugar cubes, children's blocks, or the actual construction of a cellular model using gelatin and other materials representing organelles. Using observations and models, students should compare cells with and cells without cell walls. Single-edge razor blade sections of a soft, herbaceous plant such as *Coleus* , with a generalized stain, show students different cell shapes and suggest the role of walls and wall structure. A second, and difficult, textbook-type assertion is that a chicken egg is a cell. Teachers could challenge students to devise ways to test this assertion. The exercise should demonstrate that the egg shell is not a cell wall.

Patterns of structure at the organismal level: Students should consider patterns of structure above the cellular level. Activities should explore animal body patterns and the fundamental nature of the plant axis and symmetry and should demonstrate that such patterns are not unique to, but characteristic of living things. With some teacher guidance, students can become familiar with the elements of classification by comparing body plans from representative animal groups, observing segmentation and legs, and noting differences in plant axis patterns, roots, stems, and nodes.

Interesting insights into patterns of organisms can be obtained by examining the organism in space and time. This is not, in Grades 6-8, the study of life cycles, but the study of phenomena such as proportion, scaling, and organ expansion. Following the growth of a bean plant from germination to secondary leaf formation is a useful activity for demonstrating these phenomena.

Interaction With the Environment

Response to stimuli: Living things interact with their physical environment. They respond to stimuli through tropisms and taxes and/or interact with physical factors, such as water and temperature, homeostatically. Teachers should plan descriptive activities that address "how" organisms interact with their environment and suggest "why" such interactions take place. Phototaxis in *Drosophila* or *Euglena*, geotropism in bean roots, and thigmotropism in *Mimosa* are examples of appropriate activities.

Reproduction

Life cycles: The reproductive capability of organisms is a characteristic of life. When examining life cycles, students should consider the idea of life span and look at the generalized life spans for invertebrates, vertebrates, and seed plants. Some life cycle examples should exhibit significant variation from basic patterns, perhaps insect metamorphosis or avian precocity, in light of the organism's adaptive arsenal. Observing specific stages in the life cycle of an insect such as *Tenebrio* or *Drosophila* can suggest the relationship of adaptations over life span.

Patterns of reproduction: Students should study asexual and sexual patterns of reproduction, not in chromosomal detail, but rather in terms of chromosome number: haploidy, diploidy. Activities with "Fast Plants"—vegetative propagation using *Coleus* and with the haploid/diploid stages in yeast—provide evidence for diversity of reproductive patterns.

It is important that students discuss the biological advantages of reproduction patterns for individual organisms. This discussion should address the maturity of adults, seasonal influences, and relationships of physical factors.

Composition

Chemical composition: Students should examine the chemical composition of living things, but not at a biochemical level. Activities should allow them to develop a taxonomy of substances unique to life, yet show them that elemental composition is shared with the non-living world. Before moving on to other content issues, the teacher must confirm that students understand the basic biological compounds at a descriptive level—carbohydrates, fats, proteins, and key inorganic compounds. Laboratory experiences in substance identification and location can reinforce this knowledge and help students generalize about how the location of substances in organisms or parts of organisms is important.

THE LIVING ORGANISM

The living organism serves as the second biology content organizer. At the 6-8 grade level, the curriculum should emphasize the human organism.

Systems

Human organism: Activities should focus on the human organism as a representative mammal. Students can examine how humans have adapted to a multitude of environments and what specializations have resulted. The first emphasis in studying humans, then, is the human organism as an integrated biological system.

In addition to introducing humans as mammals, activities should present humans as primates: What are the characteristics of primates? Why are humans classified with them? Such questions should lead students to consider arborealism, bipedalism, and the capacity for speech and language. Sixth to eighth-graders do not need to engage human evolution, but simply think about humans in the context of other "successful" animals. This overall view prepares students to conceptualize human systems in more than a diagrammatic way.

Human systems: Students should explore the different systems—circulatory, digestive, reproductive—but not in infinitesimal detail. Looking at each system as a model of function and comparing it to other animal systems is a productive approach. From activities, students should see a correlation between healthful living and "system maintenance." Designers and teachers must avoid a mindless "march through the body systems." Coordination with topics from other disciplines results in more dynamic experiments. For example, activities could relate circulation to principles of pressure, muscles and skeletal structures to machines and gas exchange.

Cycles

Human life cycle: Students in Grades 6-8 should study the stages of human existence, with reference to material previously learned about life cycles and human systems. Activities or discussions of human development should be descriptive and not embryological. At the 6-8 grade level, curricula should emphasize the human organism, perhaps beginning with adolescence, proceeding to adulthood, aging, death, reproduction, and back through gestation, birth, infancy and childhood to adolescence. Students should consider sub-cycles that occur within the larger human life cycle.

THE BIOLOGICAL PLANET

The third biology content organizer is the biological planet.

Components

The biological planet's hierarchical organization conditions how scientists study it. Students should begin exploration with the smallest components, niche and habitat, as this approach encourages them to recognize the ecological limits placed on specific organisms. Urban examples can provide an immediate context for some students. Students can build up to definitions of population, community, ecosystem, and biome. As they move up the organization of environmental study, students should see relationships between the biological systems and non-living components of the biosphere. They should not be asked to memorize detailed components at each organizational level. "Why" an organism lives where it lives, "how" it survives there, and "what" determines extinction are questions students should explore. Populations of organisms familiar to students can be studied, such as pigeons, starlings, and house sparrows, and activities that lead students to these questions can be designed.

Several better-known activities address these subjects: What's in a square meter? What are the organisms and their niches within that square meter? Students might construct models of ecosystems, such as two-liter bottle aquaria, in order to manipulate and understand the relationships between physical factors and biological components.

Interactions

Organisms and the physical environment: At the 6-8 grade level, activities exploring the interactions of organisms and the environment should be confined to the physical environment: limiting factors, temperature, moisture, and light. These activities also should differ from earlier activities that explored how individual organisms respond to specific stimuli. Students will consider interrelationships between organisms, including humans and other organisms, in later grades.

Patterns

Patterns of matter: Patterns in the environment include the cycling of matter: water, carbon dioxide, and nitrogen. Students should examine what evidence exists for the proportions of inorganic substances and how the evidence is acquired. Activities here can integrate easily with chemistry. The topics of evaporation, condensation, and precipitation also correlate with Earth science and physics. Since consideration of pressure and temperature occurs in Grades 9-10, activities in Grades 6-8 should emphasize empirical phenomena, such as evaporation rates, temperature, and condensation, and should involve measurable volumetric changes.

PROPERTIES OF LIVING THINGS

The cell, its structure and function, hierarchical organization, and reproduction focuses the content at the 9-10 grade level.

Structures Unique to Life

Cell functions and how cells are adapted to perform them: Having been introduced to cell structure at a descriptive level, students in Grades 9-10 are prepared to examine how the cell functions and how parts of the cell have adapted to perform these functions. Without delving into the biochemistry of cell function at this point, activities can explore photosynthesis and chloroplasts, respiration and mitochondria, transport and cell membrane (and cell wall), as models, with appropriate laboratory experiences in constructing these models. Photosynthesis at a descriptive level lends itself particularly well to a historical approach, as it illustrates how a model can be constructed from single, incremental experiments. Students can begin to derive the summary equation for photosynthesis through activities that demonstrate the importance of light. Light screens on Geranium leaves, with the subsequent iodine tests for carbohydrates, provide two elements of the process. Carbonate solutions and *Elodea* plants in bright light—students count bubbles as they emerge from the cut end of the plant—illustrates another reactant (carbon dioxide) and another product (oxygen). Hence, students can develop summary equations for photosynthesis and respiration, while reserving their elaboration as biochemical cycles for Grades 11-12.

Organizational hierarchy: With the exception of unicellular organisms (or experimentally-isolated cells), cells are organized into functioning groups. These cell groups can be as simple as a colonial formation or as progressively complex as simple tissues, complex tissues, organs, systems, and entire organisms. Activities should address cell structure, cell function, and multicellularity as manifestations of the cellular property of life and not present a taxonomy of tissue and organ types. A few examples can serve this goal: xylem and phloem and transport in plants, epidermis and dermis in animals, blood as a tissue, and general leaf anatomy. Students should be challenged to define the relationships between cell number, cell diversity, tissue organization, and organismal functioning.

Interaction With the Environment

The cell: The 9-10 grade curricula addresses a cell's interface with its environment rather than its response to environment and stimuli. One activity suited for this subject is a plasmolysis experiment with red onion cells. In another activity, increases in cytoplasmic streaming could be observed in *Elodea* cells exposed to light. Students should see the relationship of cell wall, cell membrane, vacuoles, and turgidity in plants. Likewise, they should observe for themselves the relation-ship between animal cells and osmotic pressure. This latter subject lends itself to model building, perhaps using dialysis tubing. Differential permeability experi-ments can be designed, using solutions of sucrose and suspensions of starch, with students conducting iodine tests of both dialysis tube contents and the surrounding media.

Reproduction

Reproduction of cells: Cells increase in number by mitosis and cytokinesis. Activities should introduce the mechanics of mitosis and also the equational nature of the process, i.e., the perpetuation of chromosome number and genetic identity. The place where mitosis occurs in different organisms is important for students to observe so that they can propose the role of mitosis in growth and development. Since suitable material for examining mitosis is difficult to obtain, other than some time-honored examples such as onion root tips, students could be challenged to extrapolate from one material to another. For instance, students could examine prepared slides of onion root tips and record mitotic figures in the region of cell division. Students could correlate data from several days of observing and measuring root tip growth.

In class discussion, the teacher should ask students to speculate on how mitosis could aid in chromosomal analysis. The teacher should extend the discussion to include some exploration of human chromosome studies. While teachers may not be able to obtain prepared slides of human karyotypes, they can use and xerox photographs and produce idiograms. Activities should establish that chromosome number alone is not the issue: chromosomes are qualitatively different, and a genome is not just meeting a bulk requirement. Karyotype analysis of abnormal samples can support this idea, even using the cut-outs.

Activities at the 9-10 grade level should address cytokinesis for both animals and plants in sufficient detail so that students can understand such experimental procedures as the inhibition of cell division. Observing yeast budding gives students sufficient information to propose a model for animal cytokinesis, while prepared slides of onion root tip provide some information for students about cell plate formation and cytokinesis in plant cells.

Meiosis: Meiosis as a reduction process explains the continuity of an organism's chromosome number from generation to generation and also forms the basis for the phenomenon of Mendelian independent assortment. Students need to grasp the details of meiosis not only for these reasons, but to understand the formation of gametes in animals and meiospores in plants as well. Again, working some simple non-disjunction problems that lead to aberrant human chromosomal numbers can illustrate the qualitative differences in chromosomes. Working these problems prepares students for understanding X-linked inheritance, since non-disjunction in X and Y chromosomes, using X-linked markers, helps support the important correlation of "genes on chromosomes."

Inheritance: At the 9-10 grade level, the study of inheritance should be primarily Mendelian genetics and its extensions: sex-linked transmission, linkage, and multiple allelism. These subjects provide an excellent opportunity, beginning with Mendelian principles of monohybrid and dihybrid inheritance, dominance, and recessiveness, to build an experimental case for the chromosome theory of inheritance, i.e., that genes are on chromosomes and follow patterns of chromosome transmission. Using maize seedling mutants and *Drosophila* , students can trace the intellectual development of genetics from 1865 to about 1940. Activities that make extensions to human genetics are particularly appropriate and interesting to students. Biochemical genetics and molecular biology form a large portion of the content coverage for Grades 11-12 and should be deferred until that time.

Composition

Coverage of the composition of living things is limited, at Grades 9 and 10, to how the classes of substances studied earlier, such as proteins and lipids and cell membranes, and carbohydrates and cell walls, relate to cell function.

While simple examples of enzyme action could be introduced in earlier grades, such as the action of barley seedling diastase on starch, the study of additional examples of enzyme activity on biological polymers is appropriate at this level.

THE LIVING ORGANISM

Kinds of organisms organize the content at the 9-10 grade level. The diversity of organisms was developed previously in the context of patterns of structure and reproduction in Grades 6-8. However, at Grades 9-10, coverage should be confined to taxonomic systems.

Systems

Diversity among organisms: In considering organisms on Earth, students should correlate diversity of organisms with diversity of habitat at a more informed level than Grades 6-8. Activities should stress organisms' structural modifications that are adaptations to specific selection pressures. Students should examine strategies for survival in this context. Homologies, specializations, and co-evolution are appropriate topics. Activities would look not only at the morphological variations within groups (Darwin's finches), but at adaptations for a common function (birds, bats, and structures for pollination in insects).

Classification: Understanding diversity involves classifying organisms into manageable groups. While students seem to grasp the mechanical principles of classification easily (such as taxonomic categories), it is more important for them to develop rational criteria for classification. They should see for themselves the irrationality of trying to classify organisms by color, by growth habit (trees and shrubs), or by habitat. Students should see the value of classification for scientific communication and for understanding phylogenetic relationships. Students should not be required or feel pressured to identify all plants and animals down to the Family level. In activities, students might try sorting out different "species" of oaks from a mixed collection of acorns or look at the confusion arising from common names such as "cedar" or "badger." The quantification of variation, begun in earlier grades with the study of proportions and scaling, is critical to students' eventual understanding of evolution.

THE BIOLOGICAL PLANET

Having studied the biological planet descriptively in Grades 6-8, students in Grades 9-10 examine the dynamics of organisms with their environments and with each other. Further, students investigate patterns of energy flow and build on their knowledge of how matter cycles in the environment.

Interactions

Interrelationships between organisms: Students should observe the interrelationships between individual organisms such as predation, symbiosis, mutualism, and parasitism. It is important that students look at ecosystems, in which all these interrelationships operate simultaneously. The role of rare and endangered species as indicators of the transformation of, or health of, ecosystems can be addressed. Oceans and species-rich tropical rain forests are appropriate examples of the Earth's remarkable biodiversity. Cross disciplinary explorations of the role of fire, climate changes, or volcanism on ecosystems over time also can be conducted here.

Activities should include exploration of a local "environment" or one established within the school. Students should observe interactions between populations and consider phenomena such as plant succession and pioneering populations.

Effects of humans on the environment: The 9-10 grade curriculum should address how individual organisms and populations of organisms alter the environment. These effects are not caused exclusively by human populations. Dutch Elm disease, killer bees, or chemical pollutants in water resources can initiate a discussion of the complex effects of humans on the environment, a recurring topic throughout secondary school science. The biological implications of human interaction can focus activities here: the effect of human populations on the extinction of other animal and plant species and the accelerated change to the environment leading to habitat loss or the creation of new habitats. Case studies, with the engagement of related ethical and social issues, are productive approaches to this topic.

Patterns

Patterns of energy flow: Activities should address autotrophism, heterotrophism, and the various forms of food production and utilization. Food chains and food webs are useful models for students to examine. While discussions of energy flow must address photosynthesis and respiration, they should not be presented as detailed biochemical processes.

PROPERTIES OF LIVING THINGS

The molecular basis of life focuses the content at the 11-12 grade level. Students have had experiences with the structural and functional properties of living systems and can now direct their attention to the biochemistry of life processes. There are many opportunities for integrating these topics with chemistry, and using integrated activities should help students see connections.

Structures Unique to Life

Biochemistry of life processes: Students in Grades 11-12 should confront the complex models of life processes. In particular, students should engage the processes of respiration, photosynthesis, and protein synthesis. When students examine protein synthesis, the teacher should help them relate their experiences and observations to finally arrive at the central "dogma" of DNA replication, transcription, and translation. Students also should consider exceptions to the central dogma, e.g., RNA-directed DNA synthesis.

Activities or discussion could contrast how scientists came to understand the three processes of photosynthesis, respiration, and protein synthesis. For instance, teachers could build on the descriptive historical approach to photosynthesis that students experienced in Grades 9-10. Students would consider the tools of the molecular biologist, such as chromatography, the use of isotopes, and the work of Calvin. On the other hand, teachers could discuss how scientists' understanding of the complexities of respiration, DNA structure and replication, and protein synthesis occurred in quantum leaps. In using such approaches, the teacher demonstrates that the scientific process exhibits different manifestations.

Reproduction

Molecular genetics: Engaging molecular biology leads students to examine DNA as the genetic material, the genetic code and its relation to mutation and its consequences, and recombinant DNA. With modest laboratory equipment, students can encounter the concept of "one gene = one enzyme," either with chromatography of *Drosophila* eye mutant pigments or with nutritional mutants in easily manipulated microorganisms. It is important that students make the connections between genetic material and genetic events.

Activities should be scheduled that focus on ethical issues such as recombinant DNA and its potential hazards, the human genome project, and issues concerning privacy.

THE LIVING ORGANISM

Organism development focuses the content at the 11-12 grade level. Students should investigate how a single diploid cell can give rise to a complex, highly-differentiated, multicellular individual organism.

Systems

Growth and development: Students should explore organism growth and development. They should consider the following questions in activities or discussion: Is it sufficient to define growth as an irreversible increase in mass and volume? Why or why not? How does one measure growth?

Differentiation: Activities should help students make connections between gene action, genetic control systems, and differentiation. Further, students should conclude that the study of differentiation is a great scientific frontier, providing many avenues for scientific exploration. Recent work on the mechanisms of sex determination in humans are particularly relevant and interesting since the gene-chromosome-control element relationships present a number of models.

THE BIOLOGICAL PLANET

Students should consider evolution as the great unifying principle of biology. They should observe that while other non-living systems are said "to evolve," these systems do not operate like biological evolution.

Up to this point, students have encountered descriptively the components of the biological planet and the interrelationships that address issues of adaptation. Evolution, the theories of its mechanism, and its various forms and patterns, should focus study in Grades 11-12.

Interactions

Evolution: Students should explore and compare theories on the origin of life, dealing with both chemical and biological evolution. In other words, they should consider both the formation of chemical compounds identified with living systems and theories that attempt to explain how species change over time. Controversy over whether species are fixed or mutable has gone on since well before Darwin, and scientists have proposed numerous mechanisms for explaining species change. In order to avoid the stultifying effects of lecturing about these ideas, or only reading about them, students should discuss and test theories about the mechanisms of evolution. For instance, the question of why Lamarck's explanation is inadequate to account for certain observations should require students' analysis, rather than the teacher's automatic dismissal of it.

Natural selection: Central to understanding evolution is understanding natural selection as its primary mechanism. Activities or discussion should consider the elements of natural selection, the accumulated evidence compiled by Darwin, and the theory's modification as scientists better understood inheritance. Other current theories of the mechanisms for evolution, such as punctuated equilibria, endobiocytosis, and neoDarwinism, should be studied in broad outline and the strengths and weaknesses of each analyzed.

Changing populations: Curricula should give some attention to population genetics and the important concept of the population as the evolving unit. The concept of gene frequency is particularly important and permits students to see the dynamic nature of variation in populations.

Patterns

Patterns of evolution: Activities should incorporate patterns of evolution. Students can consider evolutionary trends, including human evolution or the origin of seed plants, as well as variations in patterns, such as parallel evolution and modification by simplification.

CHEMISTRY SEQUENCE GRADES 6-12

Sub Topics	Grades 6-8	Grades 9-10	Grades 11-12
PROPERTIES OF MATTER			
Physical Properties	observation of physical properties intensive properties extensive properties	measurement of properties: density	
Chemical Properties	color change, temperature change, production of gas or precipitate	chemical composition by mass	
Properties of Solutions	conductivity, color, relative solubility	solubility precipitation concentration	solubility and solubility equilibrium
NATURE OF CHEMICAL CHANGE			
Inorganic, Organic, Biochemical Equations	word equations	balancing equations simple stoichiometry	complete mole concept and complete stoichiometry chemical reactions
Acid-Base Reactions	nature of acid and base solutions	reacting acid and base solutions	Bronsted acids and bases

CHEMISTRY SEQUENCE GRADES 6-12

Sub Topics	Grades 6-8	Grades 9-10	Grades 11-12
NATURE OF CHEMICAL CHANGE			
Oxidation-Reduction Reactions	combustion	redox defined oxidation of metals and reactivity of metals	electron transfer and reduction potentials
Rates of Chemical Change	rate observed	factors affecting rate equilibrium	derivation of rate laws
STRUCTURE OF MATTER			
Atoms	rationale for particulate model	the structure of the atom atomic structure and the Periodic Table	quantum model
Bonding and Geometry	molecules intermolecular bonds and phase change	valence shell model ionic bonding simple covalent bonds	metallic bond covalent and coordinate bonds isomers and allotropes

CHEMISTRY SEQUENCE GRADES 6-12

Sub Topics	Grades 6-8	Grades 9-10	Grades 11-12
ENERGY AND CHANGE			
Forms of Energy	heat, light, and electrical energy	light and flame tests	electromagnetic spectrum
Conservation of Energy and Phase Change	conservation of energy phase change	heat and temperature during phase change	
Changes Associated with Chemical Reactions	observation of exo- and endothermic reactions	heats of reaction	conservation of energy and Hess's law heat of reaction and enthalpy diagrams
Energy Alternatives	solar, wind, geothermal, and biomass	nuclear fission	nuclear fusion
MODELS FOR CHANGE			
Particulate Nature of Matter	the kinetic model applied to observable properties matter pictured in terms of the kinetic model	kinetic model applied to the behavior of gases collision theory	mathematical models and the kinetic model

PROPERTIES OF MATTER

Properties that characterize and identify matter is the first chemistry content organizer. The 6-8 grade content relies on direct observation for identifying properties of substances.

Physical Properties

Observation of physical properties: In order to learn classification criteria and eventually understand chemical change, students at the 6-8 grade level should observe the properties of the different kinds of materials that surround them in everyday life. These activities also help students to develop their observation skills, see the importance of measurement, and become excited about learning chemistry. The teacher should emphasize the value of both qualitative and quantitative information. At this level, students should rely on direct observation as the basis for answering, "How do we know?"

Intensive properties: Physical properties fall into two categories: intensive and extensive. Activities should help students come to the realization that most materials in the universe are not substances but mixtures of substances and do not have one set of intensive properties. Students should explore intensive properties of substances such as color, odor, texture, hardness, melting point, boiling point, magnetic character, and phase. It is important that students organize data into tables or charts for use in comparing and identifying solids, liquids, and gases. Students should categorize substances according to melting and boiling points. The study of intensive properties might begin with an exploration of the separation of mixtures such as filtration (sand and sugar; salt and pepper), chromatography (food coloring; ink from felt tip pens), or distillation (ink, vinegar, wood). To help students apply knowledge and test ideas, teachers should challenge students to devise methods for separating these and other mixtures.

Extensive properties: Centering activities around extensive properties such as length, volume, surface area, and mass using everyday materials helps students develop measurement skills. The focus should be on examining one property at a time. For example, students might begin an investigation of mass and volume by observing what happens when cans of diet soda and regular soda are placed in a tub of water. The investigation could include obtaining the masses of equal volumes of the different kinds of soda. Students should compare collected measurements, and the teacher should emphasize the need for careful measurement. Activities that establish the constancy of mass (but not of volume or surface area) in physical and/

or chemical changes illustrate the importance of mass. Activities should show that the volumes of substances change with temperature and pressure. Sixth to eighth grade students need to be able to evaluate which properties are most useful in determining the identity of a substance.

Chemical Properties

Students should compare properties carefully to determine when a change produces a new substance. They should observe a physical change and a chemical change for the same substance. Identifying chemical properties raises questions about what happens to matter when its properties change. Matter can undergo a variety of changes: fast and slow, dramatic and almost imperceptible, those producing heat and those requiring heat. Acquainting students with these changes demonstrates the potential for change in matter and the almost unlimited variety of substances that can be produced. Activities should allow students to observe many different, yet simple, chemical changes that result in color changes, temperature changes, and the production of gas and solids.

Properties of solutions

Solutions are unique in that the dissolving process does not form a new substance, as in a chemical change, but the resulting solution does possess its own properties. Emphasis should be on making observations rather than on memorizing vocabulary. Students should make and examine the various types of solutions (liquid in a solid, liquid in a liquid, etc.) and devise activities for separating them based on their physical properties. Students should observe how temperature, stirring, and surface area affect the rate of dissolving in a qualitative way. They also can determine the solubility of different substances in a given solvent, such as sugar and salt in water, and simultaneously the meaning of a saturated solution. They can investigate the solubility of a given solid in various solvents, such as dissolving sugar in water and in alcohol. Other properties of solutions that can be examined qualitatively are solution conductivity and concentration. It is important that students see that many solutes form colored solutions, and that the depth of color is related to the concentration of the solution. Care must be taken to avoid fostering misconceptions by linking saturation or conductivity with concentration. An appropriate laboratory activity for students to apply what they have learned about solubility and solutions is to grow crystals.

THE NATURE OF CHEMICAL CHANGE

Inorganic, Organic, Biochemical Equations

Word equations: A chemical equation is the shorthand by which scientists represent information about chemical change. Sixth to eighth grade students should first use word equations as sentences that convey information about how matter has been transformed. In simple exercises, students should identify both the starting substances and the substances produced. Students can replace names of substances with symbols and formulas at the 9-10 grade level when the concept of atoms and molecules is established.

Acid-Base Reactions

Nature of acid and base solutions: Students first should establish an operational definition of an acid and a base by observing how a number of solutions react with various substances and by classifying them according to their reactions. The preparation and testing of common household items with natural indicators (red cabbage, grape juice) is an appropriate activity. Linking these household items to acid-base reactions (for example, stomach acid with antacids) will help students realize that chemistry surrounds them and can lead to other activities that show how adding an acid to a base alters the nature of the base. Rough neutralizations can be performed and the salt extracted by evaporating the water.

Oxidation-Reduction Reactions

Combustion: Combustion is a dramatic and familiar example of an oxidation reaction. Activities should allow students to study the nature of fuels and the roles of oxygen and heat in combustion. Students should determine the products of combustion and might even analyze the level of carbon dioxide produced in auto exhaust and its effect on the environment. The teacher should challenge students to account for the apparent loss of mass when one or more of the products is gaseous. In a class discussion/demonstration, the teacher also could compare combustion to other oxidations such as the rusting of iron. The discussion could focus on the millions of dollars that are lost each year due to corrosion of the materials humans use. The definition of oxidation at these grade levels should be limited to the combination of oxygen with a substance and not include the "loss of electrons."

Rates of Chemical Change

Rate observed: Chemical changes occur at dramatically different rates. Students should observe several chemical changes that occur at different rates and then be asked to generalize about factors that affect the rate of change. An appropriate example would be the rate of oxidation reactions. Discussion might include the slow oxidation of paper (and the storage of historical documents), the rate at which corrosion takes place, and the rapid oxidation that occurs in burning. A demonstration showing the burning of a cigarette in air and in pure oxygen will illustrate the effect of concentration on reaction rate and provide the teacher with an opportunity to link chemistry to the real world by discussing no smoking signs in hospitals, especially where oxygen is used. Other activities that illustrate the effect of changing the temperature and surface area (by pulverizing) on reaction rates should be conducted. Teachers should help students see that the rate at which the dissolving occurs is not the same as the amount which dissolves. Students should identify types of substances reacted, surface area of reactants, temperature, and concentration as important factors in determining the rate of a change.

STRUCTURE OF MATTER

All matter has structure, and structure serves as the basis for the properties of matter.

Atoms

Rationale for the particulate model: The concept of atoms serves as a useful tool for helping students picture the submicroscopic structure of matter. Students should be introduced to the elements as the building blocks of matter. These elements interact with one another to form more complex structures known as molecules. Elements differ from one another because the particles of which they are composed (atoms) differ from one another. At this level, symbols for the elements are introduced appropriately. The writing and interpretation of formulas is best delayed until Grades 9 and 10.

Bonding and Geometry

Molecules: Bonds are the forces that hold particles (atoms) together in various configurations. The teacher might acquaint students with the variety of structures produced by particles bonding with each other by having them handle crystal models or ball-and-stick models.

Intermolecular bonds and phase change: Students at the 6-8 grade level should be able to conceptualize bonds breaking as matter changes from the solid to the liquid and gaseous phases. Students might compare melting or boiling points of various substances. As students observe variations in melting and boiling points of different substances, they might consider how bond strength is one factor that might contribute to the differences.

ENERGY AND CHANGE

The basic laws of thermodynamics apply to phase and chemical changes, but accounting for all the energy associated with change is difficult. Energy must be defined in easily measured terms to allow scientists to account for heat lost and heat gained in a reaction.

Forms of Energy

Heat, light, and electrical energy: Different forms of energy are associated with chemical change. Heat energy most commonly is associated with chemical reactions. Students should observe reactions that demonstrate the absorption and the production of light and electrical energy as well as heat energy.

Conservation of Energy and Phase Change

Heat and temperature are different, and many students will not distinguish between them. Students should observe the temperature of objects in their surroundings. Some still think that in a cool room, an object made of metal will have a lower temperature than one made of fabric. Students should observe the temperature change of water as it is heated to its boiling point and as it boils for several minutes. This activity should establish a relationship between temperature and heat and the distinction between them. Activities that vary amounts of water and activities using different substances can characterize heat as different from temperature.Through experiences with various substances, students should understand that a change in temperature depends on the nature of the substance, the mass of the substance, and the heat energy that is supplied to it. Additional activities that include boiling and freezing liquids other than water can encourage students to consider what happens to the heat when the temperature stops changing.

Conservation of energy: Students should participate in activities that demonstrate the conservation of heat energy. Mixing volumes of hot and cold water can show the loss and gain of equal amounts of heat in the mixing process.

Phase change observed: Students should observe phase changes closely. Students should measure temperatures in degrees Celsius and observe the constancy of the boiling and freezing points. Teachers should encourage students to account for energy in phase changes. Activities must establish that heat is absorbed during melting and vaporization and is released during the reverse processes. Students also should recognize that the melting point of a substance is a single temperature. Activities should contrast the melting of a mixture such as butter with that of a pure substance.

Changes Associated with Chemical Reactions

Observation of exothermic and endothermic reactions: Students should observe chemical reactions that involve heat. Teachers can assist students in grouping the reactions according to their exothermic and endothermic nature. Reactions observed by students should include some reactions that spontaneously produce heat and others that require heat energy. The teacher can emphasize the need to consider the overall or net change in energy: Is combustion endothermic because a match is needed to start a fire, or is it exothermic because the fire gives off heat once started? Teachers can challenge students to devise ways to measure heat in both fast and slow reactions.

Energy Alternatives

Solar, wind, geothermal, and biomass: Students at the 6-8 grade level should explore chemical sources of energy, such as batteries or burning fossil fuels, as limited to one time use and nonreplenishable. These reactions move spontaneously in the direction of releasing heat, and the products cannot be returned to the reactant state efficiently or quickly. In activities, students could collect heat using solar energy, wind, or other replenishable sources. The teacher can discuss geothermal sources of heat. Activities can examine the sun as the energy source supporting life on Earth.

MODELS FOR CHANGE

The kinetic theory is a basic theory of chemistry. Matter is composed of tiny particles which are in constant motion—exerting forces on each other and interacting at the submicroscopic level.

Particulate Nature of Matter

The kinetic model applied to observable properties: Matter is composed of tiny particles in constant motion—particles that exert attractive and repulsive forces on each other. In activities, students should compare and contrast the properties of solids, liquids, and gases. They should organize their observations to form a list of characteristics common to each phase. Teachers then should help students to picture particles in each phase. They should have the opportunity to make drawings of particles or use magnet boards to represent the particles in solids, liquids, and gases. (Avoid making distinctions between particles representing compounds and elements at this level.) Particles in motion but held rigidly in place in patterned structures is one model that can account for many properties of solids. Looser forces holding particles in place allows for greater range of motion and describes the liquid and gaseous phases. Teachers should use descriptive models to help students picture what happens as a sample is heated, melts, and boils. Students should be able to conceptualize bonds breaking as matter changes from the solid to the liquid and gaseous phases.

Matter pictured in terms of the kinetic model: Atoms can be pictured as tiny spheres, molecules as atoms attached to each other in specific combinations, crystals as atoms or molecules arranged in repeating patterns, solutions as particles of the solute dispersed among particles of the solvent at the atomic and molecular level. Students should apply the kinetic model to a variety of observable phenomena but describe it in their own words.

PROPERTIES OF MATTER

Physical Properties

Measurement of properties: Students at the 9-10 grade level should encounter density as an important physical property of matter. Properties of matter observed by students in Grades 6-8 now should be described in more precise terms and measured quantitatively. Activities should relate mass and volume, previously described simply as extensive properties, through the derivation of the term, density. The usefulness of density in identifying substances should be illustrated using everyday phenomena. For example, after the relative densities of the six types of plastics have been compared by floating samples in various liquids, their densities could be obtained by calculating simple ratios or plotting mass versus volume. This information then could be used to identify the plastic type. Students at the 9-10 grade level should determine and compare the densities of solids, liquids, and gases, and even various forms of the same substance that appear different such as pellets and strips of aluminum.

Chemical Properties

Chemical composition by mass: Chemical compounds have a definite composition by mass. Teachers can introduce the mass of a mole of substance as a useful measurement. Students should do activities in which they observe substances decomposing or combining, and they should collect mass data. The data allows students to derive the Law of Definite Composition. Teachers also should guide students in using the data to determine empirical formulas for simple compounds. Students should be provided with sufficient practice on the interpretation and writing of formulas and should be able to represent them using molecular models. Percent composition data can be used to establish an empirical formula for a simple compound. Students can use mass data to predict the mass ratio by which elements combine to form particular compounds.

Properties of Solutions

Solubility: Properties of solutions depend on the solubility of the solutes. In earlier grades, students encountered the relative solubilities of substances. Activities on the 9-10 grade level should help students derive a more quantitative description of solubility. Students should determine the solubility of gases and solids in water and note the relationship between solubility and temperature for each. For example, the students could determine the solubility of carbon dioxide in warm and cold carbonated beverages. They could construct solubility curves to compare the solubility of sugar and salt in hot and cold water.

Precipitation: Some ions precipitate to form insoluble salts. Certain ions are insoluble in combination with each other and will precipitate when solutions containing these ions are mixed. Students should perform tests to determine the identity of some of these combinations. They should organize their observations into solubility tables and identify the composition of solutions by the precipitation reactions they undergo.

Types: Students also should be introduced to different types of aqueous solutions according to their conductivity behavior. Distinctions should be made between conductors and nonconductors and also between solutions that become better or poorer conductors as they are diluted.

Concentration: To this point, determining relative concentrations has been useful for students comparing solution properties. But to make precise quantitative predictions, students need to define concentration mathematically. This can be introduced using activities in which students make solutions of different concentrations of colored substances (or mixtures such as powdered drink mix) by mass percentage. After moles have been introduced, students also can prepare solutions of different molarities and compare the usefulness of the two approaches. Students can use the relationship between depth of color and concentration (simple colorimetry) to determine the concentration of solutions.

NATURE OF CHEMICAL CHANGE

Inorganic, Organic, Biochemical Equations

Balancing equations: Before beginning work on balancing equations, students should compare the masses of reactants and products when precipitates and gases are formed to establish that matter is conserved in chemical reactions. Students then should be introduced to balancing equations by building reactant molecules and simulating a chemical reaction by rearranging the atoms to construct the molecules of the products. As students take the number of atoms used and rearrange them in a simulated chemical reaction, they must be sure that no atoms are lost, nor do any suddenly appear. Balancing an equation accounts for both the numbers of atoms used and the conservation of mass. Students should see equations as a shorthand way to describe what occurs when chemicals react. Substances reacted and substances formed, as well as energy, phase, and almost anything pertinent to describing the change, can be represented in simple symbols. It is important that students observe the reactions represented by the models and equations. As students encounter different chemical reactions, they can write each reaction as a chemical equation. With increased experience in observing chemical reactions, students can make generalizations about how certain substances react. It is important that the teacher help students organize this knowledge so the students can predict the products of important chemical reactions. Activities should include biochemical and organic reactions as well as more common inorganic reactions.

On the 9-10 grade level, the teacher might introduce the mole and the gram-formula-mass concepts so that students can explore realistic samples of substances. Students should see that the equation can provide information about the masses of reactants and products involved in the reaction and the total mass of the system under consideration.

Simple stoichiometry: Laboratory activities should present simple stoichiometry to help students predict masses of substances reacted and formed. These activities, which are based on conservation of mass in a reaction, allow students to derive the mass ratios in which substances react. Students should compare expected reaction yields with actual reaction yields.

Acid-Base Reactions

Reacting acid and base solutions: In earlier grades, students identified acid and base solutions by their characteristic properties. Now, activities definitely should involve the pH scale and the color changes of some common indicators. Since many students at this level will not understand the quantitative definition of pH in terms of the hydrogen ion concentration and logarithms, the pH scale should be presented as a way of determining the relative acidity or alkalinity of a solution. Students must see how pH can be controlled by adding acid or base to a water solution. Students could test local water samples and develop hypotheses to explain variations in the acid-base nature of these samples.

Oxidation-Reduction Reactions

Redox defined: Oxidation and reduction reactions involve the transfer of electrons from one atom to another. Combustion, observed in earlier grades, is revisited in Grades 9-10. Students now identify oxygen as an element that takes electrons from the atoms of the fuel during combustion. The gain and loss of electrons character-izes many common reactions. Scientists define oxidation and reduction in terms of electron gain and loss and denote the resulting change in charge.

Oxidation of metals and reactivity of metals: Students should compare the ability of various metals to be oxidized in solutions of other metallic salts. From observing the reactions of metals and ionic salt solutions, they can form an Activity Table listing the metals according to their ability to be oxidized. Students should test the definitions of oxidation and reduction by applying these definitions to other redox reactions, including reactivity of metals with salt solutions, composition reactions, and oxidations in organic and biochemical reactions. The class can consider the usefulness of certain metals based on their ability to resist oxidation.

Rates of Chemical Change

Factors affecting rate: Students at the 9-10 grade level can examine quantitatively the effect of concentration and temperature on reaction rate. For example, the iodine clock reaction can be performed with reactants of various concentrations or temperatures and the time of the reaction recorded when the blue color appears. Concentration versus time or temperature versus time curves can be constructed and used to reach conclusions about reaction rates.

Equilibrium: Students should do activities that illustrate that most reactions do not go to completion and that the products and reactants remain mixed in the same reaction vessel at "the end of the reaction." It is only when the products are removed from the reactants, such as in the case when a precipitate or water forms or a gas escapes, that a reaction goes "to completion." Even in the case of water and precipitate formation, if they remain present, equilibrium reactions exist. Students should recognize the difference between steady state reactions and equilibrium reactions. They should come to realize that conditions that increase the rate of one of the opposing processes alters the equilibrium state. Students should form hypotheses about how to shift an equilibrium and test their hypotheses using an equilibrium reaction for which a shift in equilibrium is observed easily. The teacher could use such activities to show how changes in the equilibrium state can be predicted using the LeChatelier Principle. Students could predict approximate changes in equilibrium concentrations and conditions by applying the LeChatelier Principle. The teacher should present additional examples from biochemistry to illustrate the dynamic nature of equilibrium. Students should use molecular models to illustrate what occurs at the molecular level in a reaction that is at equilibrium and to distinguish between rates of forward and reverse reactions being equal versus concentrations of reactants and products being equal.

STRUCTURE OF MATTER

All matter has structure, and structure serves as the basis for the properties of matter.

Atoms

The structure of the atom: The development of the atomic model is an instructive and useful lesson in how science works. Models form conceptual frameworks to organize complex phenomena into understandable forms. Although scientists can now see individual atoms of some substances, they still cannot see the structure of the atom itself so they build useful models to help explain observed behavior. Students could use "black box" activities to gain experience in constructing models and testing them against empirical observations.

Atomic structure and the Periodic Table: Teachers can develop activities around a blank Periodic Table or early attempts to organize the elements. Teachers should introduce the basic particles that make up an atom. Students should use the Periodic Table to derive information about numbers of protons, neutrons, electrons, and valence electrons in the atoms of common elements. Activities should define the relation between atomic number and atomic mass number, and students should apply it for identifying isotopes. The class might discuss the use of radioactive isotopes and the dangers of radioactivity.

Bonding and Geometry

Valence shell model: The valence shell model of the atom is useful in the study of bonding. Student work at this level should be based on the principle of the stable octet. It is useful for students to consider bonds as the result of forces between the nuclei of two or more atoms. Activities should examine the property of electronegativity and describe bonds by applying a simple set of rules regarding valence electrons and the attainment of a stable octet.

Ionic bonding: Ionic bonding can be explained in terms of electron transfer. The element of lower electronegativity loses electrons and the element of higher electronegativity gains electrons. In either case, the atom of each element achieves a stable octet. The loss and gain of electrons results in charged atoms. Students should use ion charges to predict formulas for ionic compounds. Activities should demonstrate how ions attract ions of opposite charge in all directions and form crystal structures. Students could grow crystals and observe crystalline structures under a microscope.

Simple covalent bonds: Covalent bonding can be explained by the simultaneous electrostatic attraction for electrons. Activities can describe the "sharing" of electrons as a different way of attaining the stable octet. Students should examine simple structures such as hydrogen, water, ammonia, and chlorine molecules. Examining oxygen, nitrogen, and some simple alkanes, alkenes, and alkynes can introduce double and triple bonds. Teachers can use bonds in amino acids, the peptide bond, and others as examples. Students should construct models of compounds as they study them. Teachers should emphasize regularly the relationship between structure and properties.

Polarity gives rise to bonds between molecules. Unequal attraction for shared electrons results in polarity. The polarity of the bonds in a molecule may result in a polar molecule. This characteristic of the water molecule is crucial for understanding its ability to act as a solvent, to form ice that floats, and its relatively high melting point. Teachers should make careful distinctions between bonds within a molecule and bonds between molecules.

ENERGY AND CHANGE

Forms of Energy

Light and flame tests: Throughout the study of chemistry, students should notice the role energy plays. Students should observe the identifying colors produced in flame tests of various metallic salts. Teachers might lead a discussion of the relationship between the emission of electromagnetic radiation and electrons.

Conservation of Energy and Phase Change

Heat and temperature during phase change: Temperature does not change during a phase change, yet heat is being absorbed or released constantly. Students should observe temperature changes before, during, and after phase change. Careful observation shows that during the phase change, the temperature does not change. The curve formed when temperature is plotted against time should generate questions about where the heat goes during the phase change. Students should explain the various sections of the curve using the particulate nature of matter. Additional activities should focus on measuring the latent heat associated with phase change. Students can use heats of fusion and vaporization to measure quantities of heat involved in phase changes. Through activities or discussion, students also should relate energy to bonding concepts. Students can apply the techniques used to measure the heat of fusion to measure the heat of solution for various substances. Teachers should ask students to speculate whether and/or why dissolving is an endothermic or exothermic process.

Changes Associated with Chemical Reactions

Heats of reaction: Energy is released or absorbed during chemical reactions. In lab activities, students should determine heats of reaction. If the same substance is used in a variety of experiments, students can compare heat values for different types of changes. For example, students could conduct activities with wax as its heat of fusion is very small while its heat of combustion is comparatively large.

Energy Alternatives

Nuclear fission: Nuclear fission has had a dramatic effect on our world. Activities should examine the principles of nuclear fission with clear examples of a chain reaction. Nuclear stability and radioactive isotopes are related to the mechanism of radioactive decay. Discussions should contrast nuclear fission of U(235) to chemical change. Both energy and mass changes should be compared for these two types of processes. The benefits and hazards of nuclear reactors as an energy source provide an opportunity for students to exercise some critical thinking.

MODELS FOR CHANGE

Particulate Nature of Matter

The kinetic model applied to the behavior of gases: Gas molecules are in constant motion and exist at relatively great distances from each other. Students should investigate the relationship between temperature and volume, and pressure and volume, for gases. They should use these observations to derive or confirm Charles' Law and Boyle's Law. With their data, students can test the Combined Gas Law and Avogadro's Law. Teachers should help students use the kinetic model to describe temperature and gas pressure. In this model, temperature is an indicator of the average kinetic energy of the molecules, but all molecules in the sample do not have the same kinetic energy. Gas pressure is due to the extraordinary number of collisions between the gas molecules and the walls of a container and the momentum per collision. Activities should allow students to compare the effects of changing temperature and pressure on a real sample of gas with the effects predicted by the ideal gas laws.

Collision theory: Collision theory applies the kinetic model to reactions. Molecules must collide for a reaction to occur. Activation energy can be expressed in terms of the energy needed for the reaction to occur. Existing bonds must be broken, and new bonds must be formed. Any change in conditions that makes collisions more frequent and more effective enhances the probability of reaction. Students should describe activation energy, reaction energies, rates of reactions, and simple equilibria using the collision theory.

PROPERTIES OF MATTER

Properties of Solutions

Solubility and solubility equilibrium: Students in Grades 11-12 should apply what they have learned previously about solutions and solubility to the concept of equilibrium. Saturated solutions and the precipitation of insoluble salts in terms of equilibrium can focus classroom activities. It is important for students to determine the solubility product for a salt of low solubility. They should make predictions about the concentration of ions in saturated solutions, possible precipitates, and relative solubility based on the value of solubility products. Additional activities should involve separating salts by selective precipitation. Other activities can apply solubility concepts to the precise analysis of water samples for the presence of various ions. In class discussion, students should consider the similarities between solubility and chemical equilibria.

Concentration: Students can learn to express concentration as molarity and study changes in freezing and boiling points as concentration changes.

NATURE OF CHEMICAL CHANGE

Inorganic, Organic, Biochemical Equations

Complete mole concept and complete stoichiometry: Using the mole, scientists can predict results of chemical reactions quantitatively. Students should investigate atomic and molecular weights, previously used to determine the mass of a mole of a substance, on a more conceptual level. Students should recognize that these are relative weights, based on a standard mass for C-12 as 12.000 amu. Students should distinguish actual mass from relative mass. Through activities, students should identify the relationship between the mole and the volume of a mole of molecules. The teacher should lead a discussion about the relevance of molar volume and Avogadro's hypothesis. In the chemistry laboratory, students could apply molar volume to stoichiometric relationships and practice solving a wide range of predictive problems. Whenever possible, activities should be designed to test student predictions.

Chemical reactions: On the 11-12 grade level, students should be familiar with a wide range of chemical reactions. They should review many of these reactions in lab work. The teacher should schedule some complex, but relevant, chemical reactions to challenge the students' understanding and application of chemical principles. Teachers should take examples from biochemistry and organic chemistry as well as inorganic chemistry.

Acid-Base Reactions

Bronsted acids and bases: Acid molecules have an ionizable hydrogen ion which, if removed, changes the particle to a base. Students can define acid and base solutions by the relative concentration of hydronium and hydroxide ions present in the solution. They should use the pH scale quantitatively to determine the concentration of hydronium and hydroxide ions present. In the laboratory students should perform acid-base titrations. Relevance to everyday life should be emphasized such as determining the "best" vinegar or "best" antacid through titrations. Through discussion, students should distinguish the strength of an acid and the concentration of an acid solution.

Oxidation-Reduction Reactions

Electron transfer and reduction potentials: From the perspective of electron transfer, students should experiment with a greater variety of inorganic, organic, and biochemical reactions. It is important that students identify oxidizing agents and reducing agents and apply these terms to redox reactions. Students should examine metals and their ability to act as reducing agents in electrochemical cells. They should use electrical potential as a quantitative measure of the combined ability of the substances to act as oxidizing and reducing agents. Activities can incorporate half-cell reactions and define relative half-cell potentials as compared to the standard hydrogen cell potential. The class should discuss spontaneity of a reaction as it relates to redox reactions and cell potentials. Describing and demonstrating the principles of the battery (electrochemical cell) is an important activity. Additionally, students should examine redox reactions that are not spontaneous such as the electrolysis of water. Teachers should ask students to consider what role electrical energy might play in chemical reactions.

Rates of Chemical Change

Equilibrium constants: It is appropriate to introduce equilibrium constants at this level. This is probably best done through the calculation of solubility products. The strength of acids can be discussed in terms of acid dissociation constants, and students can calculate equilibrium constants for other reactions using the law of mass action and the concentrations of the reactants and products.

STRUCTURE OF MATTER

All matter has structure, and structure serves as the basis for the properties of matter.

Atoms

Quantum model: The study of matter at the 11-12 grade level requires a more complete description of the arrangement of electrons in the atom. Students should consider the following aspects of the model: energy levels (n = 1, 2, 3...), sublevels (s, p, d, f), orbitals, and electron spin. Such a model applies quantum principles in a mechanical way. Students should observe line spectra for different elements and be able to use these observations qualitatively as evidence for the discrete energy levels in the atom.

Bonding and Geometry

The metallic bond: Students should explore the nature of the metallic bond and use it to explain the characteristic properties of metals.

Covalent and coordinate bonds: Many types of bonds exist with different models to describe them. Students can extend the application of quantum principles to bonding. Activities can take a simple valence bond approach. They can look at electron pair repulsion to determine molecular geometry. Students should try constructing models of molecules and relating their structure to the type of bonding orbitals employed in the structure. The class can discuss the bonding in organic substances produced by industry. Students also should examine substances found in living cells and the biochemical structures produced by cells.

The polarity of some covalent molecules, due to the arrangement of polar covalent bonds within the molecule, leads to a definition of the dipole moment, van der Waals forces, and hydrogen bonding. Students should examine properties such as surface tension, vapor pressure, boiling and freezing point, and viscosity. Students then should explain the variance in these properties in terms of intermolecular forces. Activities should help students relate solubility and miscibility to the polar nature and molecular mass of the molecules involved.

Isomers and allotropes: With a complete background in bonding, students should examine the many varieties of organic and biochemical structures. Isomers, shapes of molecules, allotropes, and bond angles are all important in determining the properties of substances.

ENERGY AND CHANGE

Forms of Energy

Electromagnetic spectrum: If possible, students should relate the origin of electromagnetic radiation emitted by elements to quantum concepts and the quantum model of the atom. Students can observe the hydrogen spectrum and the spectra of other elements. Activities can address the hydrogen spectrum and its relation to the Bohr model of the atom. Teachers could help students see how the Bohr model was useful in predicting the lines of the hydrogen spectrum, but was unsuccessful in other ways.

Changes Associated with Chemical Reactions

Conservation of energy and Hess's Law of Constant Heat Summation: The First Law of Thermodynamics is applied to chemical reactions through the application of Hess's Law. Students should use Hess's Law to calculate enthalpy changes (H) for simple reactions. Activities might be used to demonstrate that the heat associated with a reaction is independent of the path of reaction and is dependent only on the enthalpy of the reactants as compared to the enthalpy of the final products. Students should apply these energy concepts to biochemical processes, especially the energy relationships in metabolism.

Many energy considerations are summarized in energy diagrams which include the heat content (enthalpy) of reactants and products, activation energies (with and without catalyst), and heat of reaction (H). Activities could relate activation energy and heat of reaction to stability of substances.

Energy Alternatives

Nuclear fusion: Students should consider nuclear fusion as a hope for the future. Activities/discussion should compare the nuclear fusion process to both nuclear fission and chemical reactions. The amounts of energy produced and the masses involved in each process should be compared. Students should propose and consider the problems in controlling the fusion process and the pros and cons concerning its use.

MODELS FOR CHANGE

Particulate Nature of Matter

Mathematical models and the kinetic model: The kinetic model is useful in helping students picture the submicroscopic behavior of matter. Even though the model is not a precise quantitative predictor of chemical phenomena, it should be used to help students understand and explain observable physical phenomena. Teachers should discuss the kinetic model and identify both its utility and its limitations for understanding the behavior of matter. By the end of Grade 12, students should be able to use the model to describe substances and mixtures in their three states, dissolving, chemical, and physical change, as well as reaction rates, equilibrium, and other processes.

EARTH/SPACE SEQUENCE GRADES 6-12

Sub Topics	Grades 6-8	Grades 9-10	Grades 11-12
THE PHYSICAL PLANET			
Properties of Earth: Materials and Features	size and shape geologic time minerals and rocks sedimentary rocks oceans and continents fossils water (lakes, rivers, and ground water; clouds and precipitation) atmosphere	minerals igneous rocks metamorphic rocks soils volcanoes landforms waves currents non-renewable resources air masses, fronts, and storms	Earth's interior magnetism dating methods non-renewable resources
Solid Earth Processes: Crust and Interior	continental drift plates	volcanism earthquakes mountain building metamorphism	convection currents ocean basins continental drift and plate tectonics rock cycle
Solid Earth Processes: Surface	weathering erosion deposition lithification	mass movements glaciation	landscape evolution ice ages

EARTH/SPACE SEQUENCE GRADES 6-12

Sub Topics	Grades 6-8	Grades 9-10	Grades 11-12
THE PHYSICAL PLANET			
Biological Processes	fossilization and fossils human impact on the environment (stewardship)	fossil record human use of resources (stewardship)	evolution life in the universe
Hydrological Processes	water cycle surface water ground water water quality	stream erosion coastal erosion human intervention	sea-level fluctuations water pollution
Atmospheric Processes	precipitation wind	seasons causes of rain weather systems climate	air quality long-term climatic changes
EARTH IN SPACE			
	Earth-moon-sun system sun as an energy source lunar craters and phases of the moon tides solar system in space	solar system interplanetary bodies surface evolution of planets planetary motion	evolution of planetary environments astronomical influences on climate and biological evolution stellar evolution origin and evolution of the universe

THE PHYSICAL PLANET

The first content organizer is the physical planet—its properties, materials, and features, and the processes that shape its surface.

Properties of Earth: Materials and Features

Size and shape: Students should examine the size and shape of the Earth. Eratosthenes' experiment or a simple road map showing latitude can help students calculate the size of the Earth. Students can seek evidence to support a spherical Earth.

Geologic time: When studying the physical planet, students need to grasp the immensity of geologic time and the different rates at which geologic processes occur. Students might develop a geologic time line using five meters of adding machine tape per student or group. Using a scale of 1 meter = 1 billion years, students can place significant events in geologic history on the tape (first life forms, age of dinosaurs, age of humans, eruption of Mt. St. Helens). Students can begin to comprehend how little of geologic time humans or human-like life forms have occupied. The rate at which geologic processes occur can vary from very fast (earthquakes, floods, and volcanoes) to very slow (mountain building). On the geologic time line, students can plot the time required for various processes.

Minerals and rocks: Students should distinguish minerals and rocks. They should see that minerals are constituents of rocks and that rocks are the parent material of soil. One approach might challenge students to classify a mixture of local rocks and minerals. If common minerals are available, students could determine their physical properties such as hardness, streak, specific gravity, and how they break.

Sedimentary rocks: Students should understand that sedimentary rocks are classified mainly on the basis of their texture (sizes of the grains) and composition. Activities should involve either sedimentary rocks that occur locally or sediments which, if lithified, would become sedimentary rocks. Students can determine the grain sizes of local rocks.

Oceans and continents: Students already should be familiar with the relative distribution of oceans and continents. Oceans comprise much of the relatively thin layer of water covering the Earth's surface. Sixth to eighth grade students should distinguish fresh water and ocean water, perhaps by exploring their chemical differences. Activities should demonstrate that ocean water contains significantly more dissolved salts than fresh water. Students should understand that continents are composed primarily of sedimentary rocks and other rock types with a lower specific gravity than ocean rocks.

Fossils: Students on the 6-8 grade level should look at and touch different kinds of fossils and consider their location(s) in rock layers. These activities can convey the immensity of geologic time and suggest how living things have changed through time. The teacher might discuss how living things "create" natural resources such as fossil fuels.

Water (lakes, rivers, and ground water): Students should consider the properties of lakes, rivers, and ground water. The study of ground water is important because much of the water humans use is stored as ground water. Students probably believe that water flows underground in underground rivers and lakes and should learn about porosity and permeability.

Water (clouds and precipitation forms): Sixth to eighth grade students should observe and identify different types of clouds—cumulus, cirrus, and stratus—and examine how they form. Students should examine precipitation forms and distinguish rain, hail, and snow.

Atmosphere (air): It is important that students investigate the constituents of the atmosphere—including water vapor, nitrogen, oxygen, carbon dioxide, and ozone. Activities should show that air contains oxygen, but in a limited quantity. The "suffocating candle" activity can illustrate that the burning time of a candle in a closed jar depends on the amount of air in different-sized jars. Students also need to account for the composition and temperature structures of the atmosphere as a whole. The teacher might lead a discussion of how the presence of life on Earth has altered the atmosphere.

Solid Earth Processes: Crust and Interior

Continental drift: Plate tectonics currently serves as the unifying theme for the scientific understanding of solid Earth evolution. While the concept is complex in detail and requires familiarity with various geological phenomena, some experiential groundwork for continental drift can be laid in these early grades. Students can explore the fit of the continents through a "jigsaw puzzle" approach, with continents serving as pieces. Students can perceive that if the continents once were grouped together, some movement mechanism must exist or have existed.

Plates: Investigation of plate tectonic theory is reserved for later grades. The mechanics of plate movement can wait until students have explored volcanism, earthquakes, and convection currents. But students at this level can observe maps of ocean basins and continents that depict ocean ridges and major structural features. Students can observe and interpret that the "puzzle pieces" have natural boundaries formed by these features.

Solid Earth Processes: Surface

Weathering: Students should explore both physical weathering, such as abrasion and frost wedging, and chemical weathering, such as oxidation and solution. Activities should allow students to identify the relationship between physical and chemical processes. For example, physical weathering can increase the rock surface area which, in turn, can accelerate chemical weathering.

Students should note differences in the products of physical weathering (rocks and mineral fragments) and chemical weathering (clays and altered minerals). Such observations prepare them for understanding soils later. For now, they can understand that these weathered materials are available for transportation by water, wind, and ice.

Erosion: The process of weathering removes and transports materials to basins (oceans, lakes) where these materials accumulate and become sedimentary rocks. Students should consider erosion processes such as sheet wash on unvegetated slopes. The loss of valuable agricultural soils to water and wind erosion can be explored.

Deposition: Building on students' knowledge of weathering, activities should investigate how eroded material is transported and eventually deposited. Activities also might account for the products of deposition such as flood plains, deltas, and ocean sediments. The origin of most sedimentary rocks—accumulation in a basin of eroded sediments—again should be noted. Time sequence maps of large modern deltas, such as the Mississippi Delta, can show graphically how deposition of sediments builds land and fills basins. Cross sections of the northern Gulf of Mexico can suggest the great thickness of sedimentary deposits.

Lithification: Activities should build on students' knowledge of sediment deposition to investigate how unconsolidated sediment transforms into lithified sedimentary rocks. In these activities, the teacher might introduce, in its simplest form, the differences in the properties of various clastic sedimentary rocks—shales, sandstones, conglomerates—and how each is formed: depositional environments. Activities involving fossiliferous limestones could help students to see that not all sedimentary rocks form from the deposition and lithification of inorganic material. Some sedimentary rocks (e.g., coal and limestone) form due to organic processes and from organic debris.

Investigations should help students identify some factors that control the conversion of unconsolidated sediments to sedimentary rocks: compaction by the weight of accumulating sediments, cementation, increased temperature related to burial, and the effects of geologic time.

Biological Processes

Fossilization and fossils: Students at the 6-8 grade level should investigate how fossils are incorporated into sedimentary rocks. It is important that they review the different types of fossils and examine the various processes of fossilization. Activities might involve students "making fossils" by mimicking certain fossilization processes. Students should realize that fossilization is a relatively rare process—that the death of most organisms is not recorded in the geologic record. The teacher could ask students to speculate on the specific conditions required for fossils to be preserved.

Human impact on the environment (stewardship): Students should consider that humans value certain rocks due to their significant economic potential. A few examples include oil/gas-bearing rocks, coal, and sedimentary iron formations. Waters of oceans, lakes, streams, and ground water also are resources whose quality and availability humans protect for their own use. Students can research and/or document human use of these resources and observe the impact of human activity on the environment. Activities should demonstrate that many natural resources are limited, and wasteful use of resources significantly reduces reserves and shortens the time that humans can rely on them.

Hydrological Processes

Water cycle: At the 6-8 grade level, it is important for students to understand basic hydrologic properties and processes and see how features of the physical planet interact. Through activities, students should determine that water cycles to and from the oceans, atmosphere, and solid Earth through a series of processes: evaporation, condensation, precipitation, surface run-off, and percolation into the soil. Students might trace the path of a drop of water as it moves through the cycle, noting where it remains for the longest periods of time.

Surface water: Students should recognize that, on continents, surface water is found in lakes, streams, and rivers and in the form of ice. Lakes form where water fills a depression to its lowest point. Thus, water level is determined by the elevation or the outlet. Lakes are temporary features because they tend to fill with sediment or drain when the outlet erodes. The original depression may be formed by humans scooping out material or by natural dams. Students could study a lake in their area to determine who or what formed the original depression and to find the location of the outlet. Students could study the drainage pattern of local rivers and/or streams and determine the drainage divides.

Students might investigate how rivers—one form of surface run-off—simulta-neously erode and deposit as they flow over the Earth surface. Students also might investigate how large paved areas can prevent the percolation of water into the ground, and how soil erosion can result from the concomitant "sheet flow" of water across land surfaces.

Ground water: Students should understand that when a hole is dug in the ground, the hole fills with water to the level of the water table—the place where water occupies all the spaces between grains. They can encounter this phenomena by digging a hole on a beach: the hole will fill when the lake or ocean level is reached.

For a unit on ground water, porosity, and permeability, students could fill and pack a clear, one liter container with sand. Students predict whether they think the container can hold anything else. Next they put a small amount of water over and into the sand and reassess how much water they can add. Then they should add and measure the actual amount. Students should understand that the water resided in the spaces (pores) between particles and that it flowed, not as a river or stream, but by migration from pore to pore.

Water quality: Students should consider what criteria determine whether water is "good" or "bad." They could investigate the difference between soft and hard water or the effect of organic material in water. Some general ideas of pollution might be introduced at this level.

Atmospheric Processes

Precipitation: Students should investigate how dew, fog, and clouds form through the processes of evaporation, condensation, and precipitation. The dew point should be determined on days having different weather conditions. An additional activity might explore how smoke affects cloud, fog, and dust formation. Such an activity also could serve as a basis for discussing acid rain.

Students now should relate cloud types to weather—dark nimbus clouds as an indicator of impending precipitation, cumulus clouds as an indicator of fair weather. Activities also might encompass severe storms—how, where, and why they occur.

Wind: Students should understand that air movement causes winds. Air moves from higher pressure to lower pressure areas or from colder to warmer areas. In addition, the Earth's rotation influences winds. Because the air closer to the equator is rotating faster than the air closer to the poles, winds in the northern hemisphere are diverted to the right. In some areas, strong winds may transport sand and dust resulting in erosion and deposition.

THE EARTH IN SPACE

The second content organizer is The Earth in Space. Emphasis at the 6-8 grade level is on those extraterrestrial bodies with which students are most familiar—the sun and moon.

The Earth-moon-sun system: Students should examine the Earth-moon-sun system, perhaps first exploring the sun's properties and how it generates energy. Then they should explore the moon's properties and how it has undergone change. Finally, they should observe the interactions between the sun, moon, and Earth. Activities should show how the sun and Earth move relative to each other. Students could observe the length and movement of shadows during a day. They also could observe the changing position of the sun on the horizon at sunset (or sunrise) from week to week and discuss how this was used by people in the past to determine the date during the year. They could discuss how this information about the sun could be used to conserve energy in a house or apartment by blocking or admitting sunshine on east-, south-, or west-facing windows at different seasons. Observations or data tables could be used to study the duration of daylight through the year. These observations and data should help establish several concepts—first, that the Earth must rotate (as seen by the fact that there is a daylight and dark period each day) and second, that the Earth must revolve about the sun and that the Earth's plane of rotation must be inclined to its orbit (as seen by different, but cyclical amounts of daylight and dark). Activities should emphasize directly observable processes such as tides and phases of the moon.

The sun as an energy source: A study of the sun should focus on the sun as the principal energy source (heat and light) that fuels Earth processes. It is difficult to have students "discover" that the source of the sun's heat and light energy is the process of fusion, but it is important that they understand that the sun actually produces heat and light.

An exploration of annual temperature variations at different geographical points (the equator, their hometown or city, a point at or near the North or South Pole) can lead students to discover that the sun is the principal energy source for the Earth surface.

Lunar craters and phases of the moon: It is appropriate to begin studying the moon with its most readily apparent properties—craters and phases. Activities might include charting of the moon's phases and position relative to the setting sun on different days of the month, starting with the crescent moon. Students should discover that the moon is illuminated by the sun and that its phase depends on the direction and angular distance from the sun. Activities also could include observing the dark plains of the moon and the craters, which are visible in moderate-sized binoculars or a small telescope. The student observations could be compared with maps of the moon. Activities then could focus on how the moon has become cratered. Students could recreate moonlike, cratered surfaces by dropping projectiles into dry or wet plaster of paris and would discover that the crater's diameter depends on the diameter and speed of the impacting body. Students should note that the longer a surface has existed, the more craters have accumulated. Thus, craters roughly can measure age. Comparison of the moon with other planets, using photos, reveals that most solar system bodies have old surfaces (many craters), and only those with active geologic processes have "young" surfaces (few craters).

Tides: Students should examine how the moon affects tide formation. Students can gather and graph data on tide times and heights to understand that the Earth and moon move relative to one another. Graphing changes in tide height with time demonstrates the relationship between moon phases, moon positions, and the times of spring and neap tides.

Solar system in space: Students can be introduced to the ancient concept of "fixed stars" and the fact that planets first were identified because they move among the stars. (Fixed stars refers to the idea that stars are arranged in unchanging patterns of constellations.) Activities can include identifying famous constellations, such as Orion and the Big Dipper, and, depending on the season/year, charting a bright planet's motion against the background of the "fixed stars."

THE PHYSICAL PLANET

Content in Grades 6-8 focused on surface features and processes. Content in Grades 9-10 focuses on surface expressions of subsurface and atmospheric processes.

Properties of Earth: Materials and Features

Minerals: The number of different minerals that comprise most rocks is relatively small. Students might compare the properties and compositions of calcite, quartz, feldspar, and ferro-magnesium minerals.

Igneous rocks: Igneous rocks are rocks formed by solidification from a molten state. Students should explore igneous rock formed under different eruptive conditions. Basaltic rocks are the product of non-explosive volcanism. Dacitic and rhyolitic rocks are the product of explosive volcanism. Activities should demonstrate that igneous rocks, because they are composed of different amounts of silica dioxide, have different modes of eruption and produce different kinds of rock: black basalts, gray/tan andesites, and light-colored dacites and rhyolites. Observing igneous rock texture can help students draw conclusions about the relationship between igneous rock and cooling history.

Metamorphic rocks: Metamorphic rocks are rocks that show changes in mineral composition, texture, or structure due to heat and/or pressure and reactive chemical fluids. Students should note the variety of textures and mineral compositions of metamorphic rocks: slates, phyllites, schists, and gneisses. They should relate texture to the intensity of the temperature and pressure conditions. Low-grade metamorphism usually produces fine-grained rocks whereas high-grade metamorphism produces coarse-grained rocks.

Soils: Students should examine the properties of soil—its various constituents within given particle-size fractions and among different soil types. Students should determine for themselves that most soils contain an inorganic fraction—rocks and minerals, and an organic fraction—humus, leaves, and twigs. These activities might use inexpensive graded sieve sets and hand-held magnifiers. A soil study should include consideration of its environmental "relevance." Activities should show that processes which degrade or erode soil, whether human-induced or not, can affect plant growth and food production.

Volcanoes: Students in Grades 9-10 should examine the two volcano types— shield volcanoes, with their relatively calm eruptions at places like Hawaii and Iceland, and composite volcanoes, with their relatively explosive eruptions at places like Mt. St. Helens and Krakatoa. Activities should connect volcanic eruptions with the formation of igneous rocks.

Landforms: Many landforms, such as valleys and flood plains, are the product of surface processes. Others, such as folded mountains and volcanoes, result from subsurface or internal processes that cannot be observed directly and, to a large extent, must be inferred. At the 9-10 grade level, students should characterize these different landforms.

Waves: Students in Grades 9-10 should examine waves—what they are, how they are formed, what affects their size and frequency, and how they influence coastal areas. Waves are defined most precisely as an oscillatory movement of water manifested by an alternate rise and fall of a surface in or on the water. Through activities, students should formulate the concepts that: (1) waves are the movement of energy, not water; (2) ocean waves are formed generally by wind on the water surface; (3) the fetch over which wind blows affects the size of waves; and (4) the direction that waves strike the shoreline largely determines the extent of longshore drift and coastal erosion.

Currents: Activities on currents first need to establish the difference between currents and waves. Students can compare temperature and salinity between water masses to discover that the density differences can generate currents. Additional activities might focus on currents in the oceans and the effect of ocean floor physiography on current movements.

Non-renewable resources: Students should examine resources such as metals, coal, oil, and natural gas and consider their non-renewable nature. Explorations could address the use and consumption of metals and their worldwide distribution. Also, activities should show that exploding human population growth and sustained, extensive fossil fuel use not only could lead to global climate change, but depletes fossil fuel reserves worldwide.

Air masses, fronts, and storms: Using weather maps, students should identify the different kinds of air masses, fronts, and storms. They should note changes that occur over time.

Solid Earth Processes: Crust and Interior

Volcanism: Once students are familiar with the shape and distribution of volcanoes and the materials (lava, ash, gases) that volcanoes emit, activities can explore the processes that create volcanoes. Volcanism is a principal process by which ocean floors and some parts of continents are altered. Students can perceive that volcanism occurs in fairly restricted zones, rather than randomly, by locating volcanoes of the present and recent past on maps. Students should consider the impact of volcanic activity, whether calm or explosive, on humans. They also should become acquainted with the technology available to monitor and predict volcanic activity.

Earthquakes: Students should be introduced to earthquakes through accounts of what they "feel like," the damage they do, and where they occur. Maps showing the faults along which major earthquakes occur can guide activities. Cross sections (slides) showing displaced rocks and/or surficial materials at faults can illustrate their nature as fractures along which movement occurs.

Students also might map earthquake epicenters, either over time or using lists provided by the National Oceanographic and Atmospheric Administration (NOAA). Activities should demonstrate that, similar to volcanoes, earthquakes typically occur in well-defined, geographically-limited zones.

Mountain building: Students should observe mountains directly (if possible) or indirectly in pictures and on geographic and geologic maps. They can list common characteristics including elevation, folded and faulted rocks, and occurrence in belts or clusters. Students should consider what forces can deform rocks and elevate them. Students should consider whether earthquakes or volcanoes provide clues about forces in the Earth's crust and speculate where those forces might be concentrated. Geologic time can be revisited here to provide a framework for discussing rates of processes.

Metamorphism: Discussion of metamorphism should follow or overlap with mountain building. While not all mountains relate to the processes that form metamorphic rocks, many do. Students should observe the distorted nature of metamorphic rocks and perhaps perform analog-type experiments involving deformation. Students should infer that collisions of large rock masses must have produced metamorphic rocks. This inference later will allow students to conceptualize colliding plates, in other words, to describe a relationship between rock type and tectonic processes.

Solid Earth Processes: Surface

Mass movements: The downslope movements of large soil and rock masses can impact on human activities significantly. Landslides in mountainous areas or along steep slopes destroy highways and towns and block streams. Maps of major landslide areas such as the Gros Ventre area of Wyoming can illustrate these movements and their magnitude.

Glaciation: Ice covers the Earth's polar areas and is present at high elevations of many mountain ranges. Students should understand how snow accumulation can produce ice and how ice masses move in response to growth and gravity. Landforms produced by glaciers (V-shaped valleys, eskers, moraines) should be viewed on maps of active glaciers. Students then can examine maps of formerly glaciated areas of the United States. Students should recognize glacial landforms there and conclude that glaciers were once present. The advance and retreat of glaciers and any accompanying theories should be left to Grades 11-12.

Biological Processes

Fossil record: Sedimentary rock contains an amazing array of fossils and thus records life forms that existed during the Earth's history and in various environments. Students should discover that organisms adapted to particular environments and conclude that diversity in the rock record reflects both changes in life forms (evolution) and changes in the Earth's environments. By comparing the types of fossils present, one can determine the relative ages of rocks in different localities. Students should consider why certain life forms terminate abruptly, perhaps discussing modern theories that attribute mass extinctions to catastrophic meteor impacts.

Human use of resources (stewardship): The teacher might lead a discussion or schedule activities about the "Greenhouse effect." Some evidence suggests that the combustion of fossil fuels—the principal process by which electricity is generated—could lead to an increase in the total amount of Greenhouse gases in the atmosphere. The increased amount of Greenhouse gases could lead, in turn, to global climatic changes (warming). Global warming could accelerate glacier and polar ice caps melting and cause a subsequent rise in sea-level. The higher sea level could result in greater coastal erosion and the flooding of low-lying coastal areas (in which a large number of people live). The teacher should stress that this chain of events has not been proven to have occurred nor to be occurring. The current data is too limited and too complex for scientists to make a definitive statement.

Students need to understand that humans are dependent on a great number of non-renewable mineral resources. Humans should be concerned about the extraction and processing of mineral resources in the same manner as agricultural resources, forests, and fisheries. Examples of human dependence on mineral resources include the numerous materials required to manufacture an automobile and the numerous everyday products that utilize sulfur.

Hydrological Processes

Stream erosion: Activities focusing on the erosion, transportation, and deposition processes serve as a good introduction to the formation of landforms. Students can experiment with stream tables, using different substrates and water flow velocities to observe the relationships between these variables and the amount and type of erosion that occurs. The teacher should indicate that these processes occur much more slowly in nature than in these modelling activities, except during great floods.

Students should observe—in the field—the relation of erosion and deposition to the bends of a river or stream. Increased velocity produces erosion whereas decreased stream velocity (reduced energy) produces deposition. They should understand the nature of a flood plain.

Coastal erosion: Students should recognize that coastal erosion and deposition are natural phenomena determined by the amount of source material, wave direction, and wave energy. In some areas, the net result is continual coastal erosion. Here again, activities could involve stream tables—this time examining sediment deposition at a shoreline and subsequent redistribution along the shoreline as related to wave direction.

Human intervention: Because of the need for water or desire for a "view," humans frequently place their settlements or residences along river banks or coasts. When stream or coastal erosion threatens their homes, humans seek ways to curtail it. Students should investigate what people have done in their area to stop, or reduce, river or coastal erosion. They can assess how successful these measures have been.

Atmospheric Processes

Seasons: Students in Grades 9-10 should examine the seasons. Students should discover that day length varies with the season, as does the average daily temperature. Students then can build a more complete and sophisticated understanding of the relative positions of the sun and Earth annually than they developed in Grades 6-8. When asked why it is warmer in summer than in winter, most students suggest that the Earth is nearer to the sun in summer and farther from the sun in winter. The realization or discovery that seasons are different in the Northern and Southern hemispheres should help students see that distance from the sun is not the cause of seasons.

Students should explore how surface temperatures vary as the angle of inclination from the energy source to the surface varies, as long as the distance is constant between the energy source and the surface. Students can realize that the angle of inclination is another factor that affects the energy amount received. Students should examine the average distance of the Earth from the sun during various seasons. They should determine that even though the Earth is farther from the sun in summer (in the Northern Hemisphere), the angle of inclination between the sun and any point on Earth must be less than in the winter. This determination is based on the observations that it is warmer in summer than winter, and that it is summer in the Northern Hemisphere (warmer) when it is winter in the Southern Hemisphere (cooler).

Causes of rain: Knowing that precipitation occurs when warm air is cooled to its dew point, students should understand that cooling occurs in different ways. Rising, cooling air produces precipitation, and descending, warming air produces little precipitation. When air rises over mountains, the windward side receives rain. Students could look at the formation and location of rain shadows. Rain also is produced by convection and the interaction of warm and cold fronts.

Weather systems: Students should observe that most weather systems move from west to east across North America. They can deduce that the dominance of low pressure winter storms in the North Atlantic sets up "Nor'easters"—storms in which the predominant wind direction is from the northeast—and that the presence of a semi-stationary high pressure system near Bermuda—the Bermuda High—causes a dominant southwesterly flow of air along the Atlantic seaboard during the summer. Activities or discussion can relate these findings to coastal erosion and the relationship of wave angle incidence to beach sand movement. Students also can discover that weather systems move west to east by studying daily weather maps and charting the movement of weather systems (high and low pressure). Through these activities, students find that storms frequently are generated along or adjacent to fronts by the interaction of cold, dry air with warm, moist air.

Climate: Students should examine climate zones—tropical, polar, and middle latitude—and the climate types within each climate zone. Investigative activities show how various factors—seasonal winds, the presence of large lakes and/or mountains, and the different rates at which land and water warm and cool—are responsible for various climate zones and types.

A class discussion might be organized around the question: "How can or do humans affect the climate?" The discussion might address "The Greenhouse Effect" or introduce the notion that cities are local climate zones—that the presence of large paved areas, restricted winds, numerous cars and buses, and a relative lack of vegetation all combine to cause temperatures in cities to be warmer than the surrounding region.

THE EARTH IN SPACE

The second content organizer for Grades 9-10 is the Earth in Space.

The solar system: Students at the 9-10 grade level should move beyond the Earth-sun-moon system to examine the entire solar system and make comparisons between the Earth and other planets. Through activities, students should understand the immensity of space occupied by the solar system, the sizes of planetary bodies, and the distances between them. Students might construct a model solar system with planetary distances and sizes to the same scale. By using the same scale for planetary sizes and distances, students can understand how little of solar system space the sun and planets occupy.

Following the construction of a model solar system, the class might review the two basic ways scientists learn about other planets: light/spectra examinations and spacecraft missions. Students could review telescopic photos of planets and could examine how spectra reveal the presence of different atmospheric gases and surface minerals on them. Then, students could review close up photos of planets and could compare similarities and distances. This review would show, for example, that most solar system bodies have solid surfaces that have been cratered by eons of meteorite impact—similar to the moon as studied in Grades 6-8.

Interplanetary bodies: At this point, teachers could emphasize that the solar system contains millions of smaller, interplanetary bodies in addition to the major planets and their moons. The solar system contains asteroids, comets, and their fragments. Meteorites are fragments of collisionally broken asteroids that fall to Earth.

Inexpensive meteorite specimens could be passed around the classroom. They are interesting to students because they are actual rocks from space. In most cases, the rocks were broken off from parent bodies hundreds of millions of years ago. The pure iron meteorites are especially interesting as samples of iron cores formed inside the asteroids.

Students should understand that the moon craters and planet craters are caused by random impacts of interplanetary bodies, and that meteorites are "free samples" that provide clues to the internal composition of the Earth and planetary bodies. Students can examine aerial photos and world maps to understand that impact craters exist not only on the moon and planets, but on Earth. The class could revisit the crater discussion of Grades 6-8 to determine that, because the Earth has active erosional processes, most craters formed before a few million years ago have been obliterated.

Surface evolution of planets: In reviewing photos of other planets, students should recognize that the larger terrestrial planets, Venus and Mars, also have undergone erosion and volcanism. The erosional and volcanic geologic processes have erased older impact craters and modified Venus' and Mars' surfaces. Comparing these planets to the Earth, scientists (and perhaps the students) find some similar surface features. Dry river beds, dunes, and polar ice caps are evident on Mars, and volcanic mountains, fracture systems, and lava flows on both Mars and Venus.

Classroom discussion should emphasize that, on terrestrial planets, landscapes are ever-changing and evolving through geologic time in long, slow, or less continuous processes. To help students understand surface evolution, activities could include modeling of surface features much in the way that craters were studied in Grades 6-8. Using soil, sand, or clay surfaces, students can model the production of river channel systems, volcanoes, and tectonic fracture systems involving compression or tension.

Planetary motion: After examining the scale and contents of the solar system, students should explore the motions of the solar system and note the regularity in planetary movements. As seen from the North Celestial Pole, the planets all revolve around the sun in a counterclockwise pattern, and most rotate in the same direction. Students should note, perhaps through pictures and observations, that planetary paths diverge little from a plane defined by the Earth's orbit. The first modern scientific statements, Kepler's Laws, emerged from his study of planetary motions. Through Newton's work, the reason for Kepler's laws became known. Teachers might use this example to illustrate the evolution of a scientific idea.

THE PHYSICAL PLANET

Properties of Earth: Materials and Features

Earth's interior: Students in Grades 11-12 should investigate the Earth's interior. These investigations should use direct evidence (from observations and data), indirect evidence, and inference so that students experience model making and appreciate the tentative nature of science.

Students might construct a scale model of the Earth's interior with a crust mantle, outer core, and inner core. The teacher should discuss the evidence supporting this model:

- •physical data
- •seismic (earthquake) data
- •whole Earth chemistry
- •meteoritic data

Activities should show that the Earth consists of layers which differ in density. An exploratory activity might involve calculating the mass, volume, and overall density of the Earth. Students then measure the density of common crustal rock types—granite, andesite, basalt, schists, gneisses, sandstones, and limestones. The activity demonstrates that the whole Earth is considerably more dense than most crustal rocks (5.5 g/cm^3 vs. 2.9 g/cm^3), and students can conclude that the materials of the Earth's interior are much more dense than the crust.

To infer the composition of the Earth's interior, students might explore how shock waves of different wavelengths travel through and reflect from rock layers. Students then could construct a "layered" or "shell" model of the Earth's interior. If students draw on seismic data as evidence, they can propose that layers might be compositionally distinct. They would observe that seismic waves are propagated at different velocities in different rock types.

Students should revisit how meteorite composition provides clues about the composition of the Earth's interior. Because meteorites are so compositionally different from the Earth's crustal rocks, scientists infer that meteorites must have originated from the interior of an asteroid belt planet. Meteorite composition then suggests a make-up for the Earth's mantle and core: a mantle composed mostly of iron and magnesium silicates and a core composed mostly of metallic iron and nickel.

Magnetism: Curricula in Grades 11-12 should address the nature of the Earth's magnetic field. Students need to see the relationship between the geographic and magnetic poles. Activities might refer to when the magnetic pole changed position or the various times that the north-seeking and south-seeking poles have reversed.

Dating methods: Through activities, students should distinguish the relative time scale from the absolute time scale. They can determine limited spans of absolute time by studying varved clays or other rhythmic sedimentary deposits. Most age determinations are derived from the ratio of a radioactive element and its decay product. Shaking a box of coins (or of sugar cubes with one spray-painted side) and removing "heads-up" items at intervals illustrates the principles of half-life. This activity also suggests the difficulty of accurate dating when the amount of remaining radioactive material diminishes.

Non-renewable resources: Students in Grades 11-12 should explore further the depletion rate of non-renewable resources. They could list what fuels and Earth materials may be exhausted for future generations. What Earth materials do humans need to conserve? What alternative fuels and materials might future generations be forced to use?

Solid Earth Processes: Crust and Interior

Convection currents: To understand how continents (plates) move, students need examples of processes related to the Earth's interior structure that could cause movement. Activities involving models or heated water with coloring can help students conceptualize convection currents and cells. Teachers should guide students in relating their observations to the thermal regime of the Earth. To complement student activities, the teacher might interpret maps that show variations in the crust temperature (heat flow).

Formation of ocean basins: With map activities and, perhaps, films or videotapes, students should observe the varied ocean floor typography. Students need to reflect on how features observed at the continent-ocean basin margins might indicate forces causing ocean basin formation. Students should relate age variations in ocean floor rock to rift zones or spreading centers, earthquake depth of focus to trenches and volcanoes, and the magnetic field variations recorded in rocks to rift zones or spreading centers.

Continental drift and plate tectonics: Students in Grades 11-12 should study the historical evolution of plate tectonics theory. Students could review the "jigsaw puzzle" activity—first encountered in Grades 6-8—and again note how the Earth's major land masses "fit together." The teacher then might discuss Wegener's hypothesis and the evidence he offered to support it: matching rock types/structures, corresponding paleontologic and glacial data. The discussion should address the controversy caused by Wegener's hypothesis in the early 1990s and its rejection by most North American and European geologists. The teacher should indicate that continental drift remained a controversial hypothesis until the late 1950s and early 1960s. At that time, additional data on ocean floor physiography, gravity variations, surface heat flow variations, age differences in ocean floor rocks, and variations in the Earth's magnetic field were united under a comprehensive hypothesis—plate tectonics. The "umbrella concept" of plate tectonics evolved from the compilation, over time, of disparate observations. Through this discussion, students should appreciate how a major scientific paradigm shift can occur through advances in science and technology.

During activities and discussion, the teacher should emphasize that plate tectonics is still a hypothesis, but one now accepted by most Earth scientists. Plate tectonics explains many Earth processes and features—the shape of the continents, the origin and apparent temporary nature of ocean basins, and the location and activity of volcanoes, folded mountains, ocean floor features, rift zones, and earthquakes. Conceptualizing the lithosphere as segmented into large pieces (plates)—moving somewhat independently, but also interacting with one another—explains and relates these major Earth features and processes.

Once students can conceptualize plate tectonics, they can engage other large-scale concepts and issues. One such concept is the distribution and relationship between igneous rocks, metamorphic rocks, and tectonic features. Students already have encountered andesitic and basaltic volcanism and volcanic rocks. They should elaborate the relationship of volcanic rock type to plate tectonics. They should determine that (1) most volcanoes exist in the same restricted geographic zone as earthquakes, (2) that explosive volcanism is most often related to boundaries where plates are colliding, and (3) that basaltic volcanism is most often related to boundaries where plates are diverging. Students should be able to propose that partial melting of the upper mantle produces basalts at rift zones/spreading centers, and partial melting of the oceanic crust produces andesites where plates collide and subduct.

In an extension and/or application to the above activities, students can research how economic (ore) deposits of certain mineral resources form. Students would discover that plate tectonics has significant bearing on how geologists prospect or explore for certain deposits.

Rock cycle: Students should explore the cyclical relationships between igneous, metamorphic, and sedimentary rocks—the Rock Cycle. As an introductory activity, students can note similarities in mineral assemblages between rock pairs—granites and certain metamorphic gneisses, limestones and marbles, basalts and compositionally similar metamorphic gneisses. Students might use their observations to describe the relationships not only among rock types but between rock type and tectonic setting.

Solid Earth Processes: Surface

Landscape evolution: Landscape changes that occur over time—erosion wearing down mountainous areas and transporting materials away—should be investigated by students. Comparisons of old mountain systems such as the Appalachians to young ones such as in Alaska or the Cascades reveal different shapes, elevations, and slope patterns. Students can compare rivers in the deep, V-shaped valleys of young, mountainous landscapes to the meandering streams in broad valleys of older, lower landscapes. Students should associate erosive forces, acting over long periods of time, with landscape alteration. They also should appreciate that the geologic setting (types of rocks and structures) plays an important role in determining the pattern of change and the "look" of the landscape.

Ice ages: Students should examine landforms in non-glacial areas for evidence of previous glacial activity. The teacher can relate students' prior investigations of climatic change to the evidence they propose for past glaciation. The teacher might ask students, "What precipitation types and temperature changes probably are necessary for widespread glaciation?" Students can examine radiometric dating evidence for the duration of glacial stages and landform ages.

Biological Processes

Evolution: In considerable detail, the 11-12 grade curriculum should examine temporal evolution (evolution over time). Introductory activities might involve identifying differences in the same or very similar life forms over time, as recorded in the fossil record. Appropriate examples include the horse species *Equus*, the invertebrate brachiopods, and organisms from the plant kingdom. Students should engage the concept of punctuated equilibrium: evolution occurs over long time periods—not as steady gradual change, but as long periods of little change followed by short periods of fairly significant change.

It is important that students consider different evolution theories and evaluate the evidence that supports or refutes each one. Adaptation, mutation, survival of the fittest, and Darwin's ideas and evidence can be examined. Students should look at how these theories account for the extinction of life forms both in the distant past, such as dinosaurs and trilobites, and the recent past, such as the carrier pigeon.

Life in the universe: Students can approach the fascinating question of whether life currently exists or ever existed anywhere else in the universe by studying other planets, stars, and solar systems. Students might assess which planets, if any, have the capacity to sustain life. They would identify the necessary conditions for creating life from inorganic chemical compounds and for sustaining life with a supply of energy, food, and water.

Hydrological Processes

Sea-level fluctuations: Students should examine how sea-level rise affects coastal erosion. Coastal erosion, while exacerbated by human activity, is influenced primarily by events prior to human existence and over which humans have minimal control. Through the use of computer and CD-ROM-based scenarios, students could determine the amount of land that would be inundated by a one meter rise in sea-level. Then, with information about population concentration at or near coast lines, students could discover graphically the impact of sea-level rise on humans. What factors might cause sea-level rise? In addition to changes in water volume, the land of some geographic areas either is slowly sinking or rising. This phenomenon also affects coast lines.

Water pollution: Building on students' understanding of how water moves on and beneath the land surface, activities can explore water pollution in local areas. These activities might include performing perk tests on local soils or making chemical tests for contaminants in local streams, rivers, or lakes.

Atmospheric Processes

Air quality: Students should investigate how human activity influences air quality. They might study particles that settle out from the air or pollutants that are found in precipitation (acid rain) and in gases. What are the sources of pollution, and how might they be controlled?

Long-term climatic changes: Curricula for Grades 11-12 should address general, non-human induced, climate changes in the past, including sea-level fluctuations and the Ice Ages. Students could explore the temporal relationships between Ice Age evidence (glacial scour, moraines) and certain chemical indicators of paleoclimatic parameters (especially temperature) to comprehend that the Earth's climate has changed periodically over geologic time. Again, they should see that human events exaggerate on-going climatic processes.

THE EARTH IN SPACE

Evolution of planetary environments: Modern space-based techniques allow scientists to make maps of the surface and climate properties of entire planets. These maps include familiar weather maps as well as maps of mineral distribution, temperature, and ozone distribution in the high atmosphere. Classroom discussion could include techniques of remote sensing that isolate different wavelengths of the spectrum.

Activities should include plotting the measurements of carbon dioxide content of the Earth's atmosphere against time using the measures made over many years (the Keeling curve). The result shows that humans are changing the composition of the Earth's atmosphere by releasing CO^2 from burning fossil fuels. At this point, the "Greenhouse effect" could be discussed. A comparison with the planet Venus is instructive: Venus has a thick CO^2 atmosphere, and the resulting greenhouse effect has produced surface temperatures approaching 900° F. Without a greenhouse effect, the offsetting combination of lower solar distance and more reflective clouds would lead to Venus temperatures approximating those of Earth.

The above discussion provides an opportunity to discuss the relation between science and public policy. For example, no single industry or nation can solve CO^2 or ozone problems. The activities also offer opportunities for coordination with other science disciplines. The CO^2 greenhouse effect involves physics of light absorption, chemistry of CO^2 production, and biological processing of CO^2 and carbon in plants.

Astronomical influences on climate and biological evolution: Although the slow and steady processes of familiar geologic evolution arise from within the Earth's environmental system, a few dramatic effects probably arise from extraterrestrial influences. For example, some theories of Ice Age sequences invoke long-term, small, cyclic changes in the Earth's orbit.

Another likely dramatic effect was the impact of a moderately large asteroid 65 million years ago—the time when the existence of large reptiles ended abruptly. The class could consider recent evidence for the asteroid event including the contents of the thin soil layer marking the end of the Cretaceous period. These contents include enhanced iridium from asteroid material, spherules of fused glass produced from melting of rock and soil, shocked quartz grains, tsunami deposits in coastal regions, and enough soot to represent the burning down of all the world's forests.

Examining the likelihood and evidence for the asteroid event could lead the class into a discussion of how scientific ideas evolve. "Catastrophism" in pre-1800 geology was replaced by the "uniformitarian" ideas of slow and steady geologic evolution. But current scientists suspect that occasional climate catastrophes, such as asteroid impacts, may have been superimposed on uniformitarian climatic evolution. If so, the puzzling mass extinctions in the biological record make more sense.

Finally, students might look at public policy issues related to asteroids, i.e., scientific proposals to "blow up" approaching asteroids (to protect the Earth from impacts) or those that recommend space explorations to "harvest" asteroid resources (metals).

Stellar evolution: Students should become familiar with extra-solar bodies—the stars. Students should recall their observations of stars in constellations in Grades 6-8, and then look at solar and stellar spectra. They should conclude that stars look like the sun. The class could consider the question, How much farther away would the sun have to be to look as faint as a star? Curricula should acquaint students with stellar brightness, masses, sizes, distances, and varieties such as sunlike stars, red giants, and white dwarfs. The latter represent different stages of evolution relative to the amount of hydrogen that has been consumed by fusing into heavier elements.

Spectral analysis can establish that a relatively small number of elements compose the universe. Applying Wein's and Stefan's laws helps explain colors of stars, and, in turn, their temperatures. Students may be able to make calculations of masses and sizes indirectly using other scientific laws. Activities could include parallax measurement as a tool for determining stellar distance, as Tycho Brahe did in the 1500s using simple observations.

Origin of solar system and other stars: By referring to chemical experiments, students should recognize that as gases cool in the universe, different solid particles or liquid droplets begin to condense out of the gas. Raindrops, snowflakes, hailstones, and particles of smoke are examples. Meteorite evidence suggests that such particles condensed within cooling gas around the sun after it formed. Just as falling snowflakes sometimes collide and stick together, these particles collided and aggregated. This example could be used to introduce the modern theory of planet formation—the accretion of many dust particles into a few larger bodies during collisions in the gas or solar nebula around the newly formed sun.

The teacher should review the principles of radiogenic dating for students as well as the ages of various types of meteorites and lunar rock samples. This information reveals that all available planetary materials, from different parts of the solar system, formed 4,550,000,000 years ago during a fairly short interval of about 20 to 50 million years.

Origin and evolution of the universe: Slipher and Hubble made a major discovery of this century—that virtually all galaxies appear to be moving away from Earth. (From this observation, students should not conclude that Earth has a special place in the universe; scientists explain the phenomenon as the expansion of time/space). To study the radial motion of objects at great distances from Earth, students can apply known terrestrial laws, for example, the Doppler Effect to light. Students can arrive at Hubble's law through such applications coupled with measurements of galactic distances (also by indirect means). Students can observe and then mentally reverse the expansion process to realize that, in the past, galaxies were closer together. Students then can extrapolate back in time and propose one theory on the origin of the universe—the Big Bang. They should consider other explanations.

PHYSICS SEQUENCE GRADES 6-12

Sub Topics	Grades 6-8	Grades 9-10	Grades 11-12
MATTER			
Kinds and Characteristics	solids, liquids, and gases		the electron

metals

the atomic nucleus |
| **Properties** | mass

volume

density

temperature | pressure | resistivity

the size of atoms and nuclei

radioactive half-life |
| **Change Processes** | melting and boiling | pressure vs. temperature at constant volume

volume vs. temperature at constant pressure | ionization

radioactivity |
| **Models** | | kinetic theory model | the atomic model |
| **MOTION AND FORCE** | | | |
| **Descriptors of motion** | distance traveled

time for trip

average trip speed

direction of motion | average velocity over short time intervals

accleration | |

PHYSICS SEQUENCE GRADES 6-12

Sub Topics	Grades 6-8	Grades 9-10	Grades 11-12
MOTION AND FORCE			
Causes of motion	Newton's first and second laws (and friction)	Newton's second law (in one dimension) velocities and force as directed quantities Newton's second law (in two dimensions but qualitative) circular motion	vector addition Newton's second law
Equilibrium	simple machines	drag force and terminal velocity	stability of nuclei nuclear forces
Momentum: A Conserved Quantity		conservation of momentum	conservation of momentum
ENERGY			
Kinds of Energy	kinetic thermal work potential	potential energy in springs and magnets energy of batteries	radiant energy
Energy Transform-ations		kinetic <-> potential mechanical <-> potential battery energy to thermal energy	fission fusion

PHYSICS SEQUENCE GRADES 6-12

Sub Topics	Grades 6-8	Grades 9-10	Grades 11-12
ENERGY			
Conservation of Energy	simple machines	specific heats	quantized energy levels conservation of mechanical energy mechanical equivalent of heat first law of thermodynamics
Second Law of Thermodynamics	convection conduction		heat engines and refrigerators
ELECTRICITY AND MAGNETISM			
Static Charge	attraction and repulsion two kinds of charge neutralize each other charge is conserved		Coulomb's law
Moving Charge and Magnets	sparks and lightning conductors and nonconductors compass needle near permanent magnet compass needle near a current	ammeters and voltmeters Faraday induction	magnetic forces on moving charges magnetic forces on currents

PHYSICS SEQUENCE GRADES 6-12

Sub Topics	Grades 6-8	Grades 9-10	Grades 11-12
ELECTRICITY AND MAGNETISM			
Electric Circuits	DC circuits must be complete current in a series circuit	Ohm's law Power law	parallel circuits
Fields			static and magnetic fields
WAVES AND LIGHT			
Mechanical Waves	water waves and wave fronts waves and pulses sources of waves	interference of water waves	
Sound	sources of sound pitch and loudness	interference	Doppler effect
Light: Characteristics and Models	energy transfer by heat radiation	refraction total internal reflection interference wave model of light	light as an electromagnetic wave the photon model of light merging the models

MATTER

Kinds and Characteristics

Solids, liquids, and gases: To become familiar with the physical nature of matter, students should encounter common examples of solids, liquids, and gases. Wood, pure metals, alloys, rocks, salt and sugar crystals are appropriate examples for solids. Water, vegetable and mineral oils, honey and sugar syrups, and rubbing alcohol are suitable for liquids. Examples of easily observable gases are more difficult to find, but air, steam from boiling water, and carbon dioxide as it sublimes from dry ice work well.

Properties

Mass: In order to compare differing amounts of matter, students must be able to measure mass. Students in Grades 6-8 should become comfortable using the tools for measuring mass qualitatively, such as equal arm balances, beakers, and graduated cylinders. These tools must be sensitive within the range of measurement required for each activity, especially when students try to make mass measurements of a gas.

Volume: Volume describes the amount of space that matter occupies. To learn about volume, students might construct regular solids from uniform building blocks, measure volumes of liquids poured in a calibrated beaker, and determine the volumes of irregular solids immersed in this beaker by measuring the volume of the displaced liquid. Malleable but incompressible solids, such as putty or water-insoluble clays, can illustrate that some solids have the same volume regardless of shape. Compressible solids, such as plastic foam, show a decreased volume when they are compressed. Students can build simple, mechanical models of gases to explore relationships among the amount of gas present, the container's shape, and the results of simple volume compression or expansion.

Density: Students should consider density as a property of matter. Students might, for example, determine the mass and volume of different-sized pieces of the same rock and find the ratio of mass to volume. If mass versus volume is plotted on a graph, the slope of the graphed line is the rock's density. Activities should allow students to compare densities. Using the previous example, students could find the densities of pieces of iron or steel pieces and compare them to that of the rock.

Temperature: Temperature activities at the 6-8 grade level should involve measuring the hotness or coldness of matter with a standard student thermometer.

Activities/discussion could address students' preconceptions concerning heat and temperature. An object that "feels" hotter may not have the higher temperature. Students also need to be informed that heat conductivity effects might confuse their temperature measurements of different materials.

Change Processes

Melting and boiling: Students should observe the phase changes of water from solid to liquid, and subsequently, liquid to gas. They should be encouraged to ask questions about heat and temperature and their effects on the system. Students should investigate phase changes while recording temperature changes in equal time intervals as heat is constantly added to the system. This activity enables students to correlate the constancy of the system temperature as ice melts and water boils as something that occurs within the same phase. After graphing and analyzing the data, students can see that temperature increases as heat is added unless two phases exist simultaneously. They should note that 0°C is both the freezing and melting temperature, and 100°C is both the condensing and boiling temperature. Students also should discover that some phase changes are reversible, perhaps by subtracting heat from the ice-water-steam system.

MOTION AND FORCE

Descriptors of Motion

Distance traveled: Teachers can introduce distance traveled as the amount a car odometer marks during a trip. Activities could have students measure distance traveled along prescribed paths through the laboratory or school grounds using a tape measure or short ruler. Students should consider how backing up adds to the distance traveled.

Time for trip: Sixth to eighth grade students, who have kept track of how long it takes to go from home to school or from the airport to a distant city, already should understand time for trip. Students should consider whether time at rest counts toward trip time.

Average trip speed: Average trip speed equals distance traveled divided by time for trip. Students should measure "distance traveled" and the associated "time for trip" for several real trips. From this experience they should realize that average trip speed is higher (for the same trip) if the traveler hurries.

Direction of motion: Through activities, students should conclude that when an object moves, it moves in some direction. They also should note that the direction need not stay constant as motion continues.

Causes of Motion

Newton's first and second laws (and friction): Students easily comprehend that if nothing is done to an object at rest, it remains at rest. They do have difficulty accepting that if nothing is done to a moving object, its motion continues un-changed. Friction causes most moving objects to slow down, and students have difficulty seeing friction as "doing something to the moving object." Activities should allow students to play with a system that closely approximates no net force and also with a system with an unbalanced force. These activities should help students verbalize and draw pictures of forces that act on objects to produce constant and accelerated motions. In class forums, students can devise "force measurers" and define the magnitude and direction of forces in common terms. They should begin to view forces as agents that engender changes in an object's type, speed, or direction of motion. Students also might conduct further investigations with friction.

Equilibrium

Simple machines: Students should discover that with a machine they can raise a massive object using much less force than they would need to lift the object straight up. Activities should utilize levers and inclined planes and student-innovated force detectors. Students should determine by what factor the force exerted multiplies to lift the object and measure the distance through which the force acted and the distance the object was raised. Accuracy is not critical to these activities, but students should recognize that the ratio of the forces is comparable to the ratio of the two distances. Students will build on these activities when later studying conservation of energy.

ENERGY

Kinds of Energy

Kinetic: Curricula should include experiences involving different kinds of energy. Students need a simple operational criterion to identify when an object has acquired or expended energy. Teachers could, for example, assert that whenever something moves a box across a rough table top, that the something expended energy in the process. Thus, when a moving cart hits the box and pushes it forward, the box expends some kind of energy. A clue that the cart gave up energy is that now it is moving more slowly. With this example, students should see that motion enabled the cart to push the box and conclude that moving objects have energy. They should define this "moving" energy as kinetic energy.

Thermal energy: Building on the previous criterion, students might investigate hot objects as objects possessing energy. Activities could involve a steam engine—a device that takes in heat and causes objects to move. A simpler activity could entail boiling water in a flask, allowing the steam to flow through a hose and directing it against a paper box. The students may say that the gas' motion moved the box, i.e., kinetic energy, and not observe the thermal energy of the boiling water. Yet this is a critical observation, and students should be encouraged to explore energy as it changes from one form to another.

Work: When a force acts through a distance on a moving object, the force does work on the object. If students push a box along the table, they do work on the box and may increase the box's kinetic energy if its speed increases. If friction exists between the box and the tabletop, both surfaces will be warmed by the rubbing. Students should notice that thermal and mechanical energies were produced here at the expense of their personal chemical energies. Students might enjoy constructing mini-projects that build on the correlations between work and mechanical energies and biological energy expenditures.

Potential energy: Activities should examine how objects can possess energy by virtue of their position. For example, a student could place a cart on a platform above a table top and let it roll down a ramp onto the table. If the cart now hits a box, the box will move, so the cart must have had energy even when at rest on the platform. Students should define this energy as potential energy. The potential energy was converted to kinetic energy as the cart rolled down the ramp. The cart acquired potential energy when someone did the work required to lift it to the platform. The work overcame the influence of gravity, so students should understand this energy (cart's energy at rest at the top of the ramp) as "gravitational potential energy."

Energy Transformations

Students should observe energy transformations directly as changes in mechanical energy from potential to kinetic energy and vice versa and the loss of mechanical energy through friction. This transformation into thermal energy can be explored by discussing the braking of a moving car or by demonstrating a rotating bicycle tire brought to rest by pressing a pad against it. Coordination with other science content areas or incorporation of science and technology topics works well here.

Conservation of Energy

Simple machines: Activities should introduce the idea that the work (in joules) done by a force acting through a distance is the product of the force (in newtons) and the distance (in meters). Activities also should suggest that the amount of work is equal to the energy transferred from the agent to the object on which work was done.

Students could return to the earlier activities with levers and inclined planes. They could use a spring balance calibrated in newtons to measure how much force it takes to lift some moderately heavy object. They lift the object, while it is on the short side of the lever, by pulling down on the long side with the spring balance and record the force required. They measure the distance the object went up and the distance the spring balance went down. Then, students could calculate the work done by the person who pulled down the spring balance and the potential energy increase of the lifted object. A comparison of these two quantities should reveal that they are nearly equal. Finally, students could perform a similar activity with an inclined plane. This time, work done on the object is the force necessary to drag it up the incline times the distance, measured parallel to the incline, that the object moved. The increase of potential energy of the object is, as before, the weight of the object (in newtons) times the vertical elevation increase (in meters).

Second Law of Thermodynamics

Convection: Through activities, students should identify that unless work is done by an external agent, heat always flows from matter at a higher temperature to matter at a lower temperature. They should understand that heat transfer can occur through different processes, and they should experiment specifically with convection—the flow of hot matter into its cooler surroundings.

Conduction: Conduction can be demonstrated with students holding one end of a metal rod and inserting the other end into an open flame. After a few minutes, the cool end will begin to feel warmer. Clearly heat passed through the rod, but matter did not. Students should define this phenomenon as conduction.

ELECTRICITY AND MAGNETISM

Static Charge

Attraction and repulsion: Activities should examine how charged objects exert forces on one another. Rubbing rods with cloth or fur and bringing them near hanging pith balls is a simple example. Students should encounter examples of both attraction and repulsion. Teachers should encourage students to speculate on a model consistent with attraction and repulsion observations. Students need to consider the following questions: 1) What is the mechanism that produces the charge? 2) How many different kinds of charge are there? 3) What are the rules that determine when force is attractive and when repulsive?

Two kinds of charge neutralize each other: Activities should demonstrate that two kinds of charge neutralize each other. A possible activity involves two pith balls suspended from strings. Students charge one with positive charge and one with negative charge. They record the approximate strength of the repulsion when a positive rod is near the positive ball and a negative rod is near the negative ball. They use as a measure of strength the string's angle with respect to the vertical when the rod end is held in a predetermined location. Students remove the rods and touch the two balls to each other. Students should conclude from the resulting symmetry that both have identical charges. They must determine which rod has the same kind of charge as the balls and put it at the same predetermined place. The force of repulsion will now be less than before, suggesting that the two kinds of charge neutralized when the balls touched.

Charge is conserved: Students probably cannot determine this principle quantitatively. The teacher can suggest it, however, by having the students observe that whenever they produce charge of one kind on a rod by rubbing it with a cloth, they always produce charge of the other kind on the cloth. Thus the students did not create a net charge by rubbing; the rubbing simply rearranged charges that already existed.

Moving Charge and Magnets

Sparks and lightning: Students should examine how large amounts of charge on one object can suddenly jump to a nearby object. When this jump occurs, light is produced. Two garments that exhibit static cling make easily visible sparks when separated in a dark place such as under a blanket. Probably every student will have felt the spark that goes from hand to nearby metal door knob after walking across a thick wool or nylon rug. Students may not have realized that lightning is an electric charge jumping from one turbulent cloud to another or to the ground.

Conductors and nonconductors: Students should observe conductors and nonconductors. The teacher could use a 1.5 volt dry cell for a demonstration. First, the teacher produces a spark by touching one end of a wire to one electrode and bringing the other end of the wire to the point where it lightly touches the other electrode. Seeing the spark should convince the students that the two electrodes have opposite kinds of charge on them. The teacher then connects one metal wire from each electrode to a terminal of a flashlight bulb. The bulb lights. Then the teacher replaces the metal wires with cotton thread. The bulb does not light. Students see that metal conducts charge, but cotton does not.

Compass needle near permanent magnet: A compass needle is a permanent magnet, and students are familiar with the idea that it aligns in a north-south orientation because the Earth is a magnet. Students can experiment with magnets: they could bring a compass needle near an iron bar magnet in order to establish that the bar is a magnet that has more influence on the needle at short range than the Earth does. Students discover that the two ends of the magnet behave differently: one attracts the north pole of the compass needle, and the other repels it.

Compass needle near a current: Students should continue experiments with magnets. They could bring the compass needle near a wire that is conducting a charge (as shown by a lighted bulb in the circuit). The deflection of the compass needle shows that, similar to a permanent magnet, an electric current produces magnetic effects in its vicinity.

Electric Circuits

DC circuits must be complete: Students on the 6-8 grade level should discover that charge can move in a complete circuital path. Activities can center around constructing circuits with batteries. The teacher can give each group of students a battery, a bulb, some wires, and clips that help them make connections. They should try to light the bulb with as many different arrangements as they can discover. Class discussion can lead the students to conclude that successful arrangements provided a path from one terminal of the battery to one terminal of the bulb, then through the bulb filament, out the other terminal of the bulb, and to the other terminal of the battery. If they can imagine that a charge also can move through the battery internally, they must conclude that a complete circuital path exists.

Current in a series circuit: Activities should disprove students' frequent misconception that charge is used up as it moves through a circuit. Students could connect two identical light bulbs in series with a battery. They both shine equally bright—showing that each has the same amount of charge flowing through each second. Students then insert a resistor into the circuit—one large enough to dim the bulbs but not extinguish them. Students should conclude that no matter where the insertion occurs, the bulbs behave the same way.

WAVES AND LIGHT

Mechanical Waves

Water waves and wave fronts: Students should be provided with opportunities to observe waves carefully. Water waves are familiar and thus serve as a good introduction. When ripple tanks are placed on top of an overhead projector, the waves can be projected on the wall. Much can be done with nothing more than a large, shallow pan, and student groups can work with individual ripple tanks. Students can tap the surface once and watch the wave pulse travel away from the point of disturbance. In particular, students should note the circular wave front. Students then can use a ruler to create a pulse with a plane wave front. Finally, they can tap at the midpoint periodically, thereby creating a continuous train of circular waves. Students should observe that when the tapping frequency is high, the distance between successive crests (wavelength) is low, and vice versa.

Waves and pulses: Students should observe additional examples of pulses and waves. Using a long rope tied to a fixed point at one end and held by a wave maker at the other, students can examine wave speed and what it depends on. Students can watch a pulse go down the rope. The teacher could change the tension in the rope or the amplitude and see if the students agree that the speed increases with tension but is almost completely independent of amplitude. Students also may note that the pulse reflects from the fixed end upside-down.

A slinky, suspended horizontally using string, can illustrate a longitudinal wave. The fact that the oscillations are in the same direction as the wave propagation makes these waves look very different, but they have the same general features: frequency, wavelength, and amplitude.

Sources of waves: Students should understand that waves spread out from a disturbance in a medium. In addition, some kind of restoring factor exists that causes an oscillation. Because every part of the medium is linked in some way to neighboring parts, the oscillation travels through the medium at a speed that depends on how tightly neighboring parts are coupled and how massive the oscillating parts are. Activities could demonstrate this mass dependence using masses hung from several equally spaced points on a slinky.

Sound

Sources of sound: Students in Grades 6-8 should think about the sources of sound. Students could use tuning forks to make sounds. They should understand that energy was given to the tuning fork when it was struck, and conclude that their eardrums oscillated, so energy traveled from the tuning fork to their ears. Students should speculate whether matter carried this energy or a wave transmitted it.

Pitch and loudness: Students should associate pitch with the frequency of the oscillating source and loudness with the amplitude of that oscillation. Tuning fork activities should convey the idea that if one strikes a fork harder, the amplitude of oscillation is larger, and the note sounds louder.

Light: Characteristics and Models

Energy transfer by heat radiation: Students should explore how energy travels from a source to a detector. Activities could draw connections to the convection process. Activities might involve a high wattage light bulb or a red-hot filament such as one finds in a toaster. Students feel the heat while standing some distance from the source. Students should speculate on how the energy reached them. If they suggest convection, the teacher should point out that the sun also sends out heat, but surely not by convection. Students should consider whether and how the two processes could be similar.

MATTER

Properties

Pressure: Ninth to tenth graders probably think of pressure as another word for force. The teacher can schedule activities to help students distinguish pressure and force. Scientists refer to gas pressure rather than gas force because the pressure is uniform throughout a modest-sized container of gas whereas the total force on any surface immersed in the gas depends on the surface area.

Change Processes

Pressure versus temperature at constant volume: Investigating pressure changes in a given volume of gas using three different temperatures will allow students to see that the plot of pressure versus temperature is linear. The three temperatures should be as far apart as possible because the spread improves the accuracy of student graphs. Water and ice (0°C), water and steam (100°C), and liquid nitrogen work well, but may not be available in a secondary school classroom. Dry ice in acetone can substitute for the lowest temperature. (If these materials are not available, teachers may have to describe the experiment rather than have students conduct it.) Students graph pressure versus temperature and extrapolate the straight line through the data points back to zero pressure to find an experimental tempera-ture for absolute zero. Students can define this new temperature scale as the Kelvin scale and compare it to known temperature scales.

Volume versus temperature at constant pressure: Students should experiment with keeping the pressure of a gas constant and observing how temperature changes cause the gas to expand or contract and its volume to increase or decrease. Schools can purchase equipment designed for these activities from companies that handle scientific apparatus.

Models

Kinetic theory model: Kinetic theory can be introduced by having students suggest a model that describes how gases behave and predicts what gases do under new conditions. One model students might propose is that gas particles are like small marbles. As the gas temperature rises, the particles move faster, and they hit the container's walls and each other harder and more frequently. Hitting the walls harder and more often increases the average force and pressure on them. This model should predict that energy, and therefore the momentum, of each particle goes to zero at absolute zero. Thus, the pressure on the walls is zero. Unfortunately, this model does not hold at very low temperatures.

Another model students might propose is that each molecule expands when the temperature increases, thereby increasing the volume of the container if its walls are movable and increasing the pressure on the walls if they are not movable. If this model were correct, students would predict that hot molecules would have more trouble passing through small holes than cold molecules, but this is not true. The important consideration here is to encourage students to propose models and to deduce whether the models they propose are consistent with observations.

MOTION AND FORCE

Descriptors of Motion

Average velocity over short time intervals: Students at Grades 9-10 already understand the concept of average trip speed. Activities should encourage them to rethink this concept letting the length of the trip get shorter and shorter. They should recognize that in a long time interval, the object could be going slow at one instant and fast at another, but in a short interval, there is little opportunity for this to happen. Students should conclude that for very short time intervals, the average trip speed provides a good measure of how fast the object moves at all times during that interval.

Acceleration: Students should explore the basic principles of acceleration and deceleration. Appropriate activities enable students to record an object's position at successive times separated by short intervals or to graph the motion of moving bodies directly on computer screens. Students either plot data for average speed versus time or study computer-generated graphs to determine if the speed is constant, increasing, or decreasing for the given time intervals. Students should study motion when the acceleration is zero and when it is a constant value.

Causes of Motion

Newton's second law (in one dimension): Students should explore how acceleration varies with the force that causes it and the mass of the object accelerated. Students can participate in computer-aided activities or more traditional activities to gather data and make graphs of acceleration versus force and acceleration versus mass. It is important for students to generate plots which indicate that acceleration is: (1) directly proportional to the force when the accelerated mass is constant and (2) inversely proportional to the accelerated mass when the force is constant. Students can combine these two relationships to conclude that acceleration is proportional to the ratio of the net force to the accelerated mass.

Velocities and forces as directed quantities: In Grades 6-8, students were introduced to motions along curved paths and simultaneous forces that acted in different directions. Students should review these concepts and extend them to include accelerations. Activities/discussion should introduce velocity as the (vector) quantity that describes both the speed and the direction of motion. Students should recognize that whenever a moving body changes direction, it changes velocity. They also should realize that, since acceleration describes the rate of change of velocity, an acceleration occurs in curved motion when the speed remains constant (because the direction is changing). Students also need to know that a change in direction requires a force.

Newton's second law (in two dimensions): It is important for students to explore the relationship between force and acceleration for two-dimensional motion. From activities, students should conclude that acceleration is proportional to the force in two dimensions, as it was in one dimension, and the net acceleration (change in velocity) is in the same direction as the net force.

Circular motion: Students should explore uniform circular motion and understand that circular motion requires a force directed toward the center of the circle. Activities could involve placing a mass on the end of a spring and swinging it in a horizontal circle. The inward force on the mass is obvious because the spring will stretch during the motion. The teacher could demonstrate other kinds of motion and ask students what provides the centripetal (inward-directed) force. For a train on a circular track, it is the track pushing inward on the wheels of the train. For a student running in a circle, it is the traction between her shoes and the floor. The teacher might introduce this concept by asking if students know what keeps satellites in circular orbits about the Earth (gravity). Students should notice that the magnitude of the force increases with the speed of the object. This is easy to show using a moving mass on a spring. The activity should pose questions about acceleration: Is there one? What is its direction? Why does the velocity change even though the speed is constant?

Equilibrium

Drag force and terminal velocity: Ninth to tenth grade students should consider drag force and how it increases with velocity and is always in the opposite direction. Activities might explore how a falling object experiencing a drag force eventually reaches a velocity such that the drag force is equal to (but opposite from) the force of gravity. When the forces add up to zero, the acceleration is zero, and the falling object maintains a constant velocity called terminal velocity. Students could observe small spheres reaching terminal velocity as they fall through viscous fluids such as mineral oil or white karo syrup.

Momentum: A Conserved Quantity

Conservation of momentum (in zero-momentum frame): Students should observe and feel the effects of two objects at rest pushing off each other. For a classroom demonstration, the teacher could align two large flatbed carts near each other and load them with different numbers of students, perhaps one student on one cart and three on another. Two students on different carts push off each other as hard as they can. The class should observe that the lower mass acquires the larger speed. Students can explore the same ideas more quantitatively using traditional spring-loaded carts and masses, rolling bowling balls at each other, or compressing a stiff spring between two wood blocks and releasing them simultaneously.

ENERGY

Kinds of Energy

Potential energy in springs and magnets: Students should explore potential energy using springs and magnets. Since activities in Grades 6-8 developed the idea that energy changes forms but does not emerge from nothing, students should consider where and how the transitions between potential and kinetic energy occur. Students should investigate elastic potential energy and also review gravitational potential energy. They also might explore energy concepts using magnetic forces: Bar magnets attached to rolling carts can be made to attract or repel—generating kinetic energy when the carts are released from rest.

Energy of batteries: Students in Grades 9-10 should examine how thermal energy can be generated by producing an electric current in a light bulb or toaster. If the current source is a battery, then students can conclude that the battery must store energy. Students can compare this energy to chemical energy because of the chemical changes that occur in the battery. But, in this case, since the energy is carried from battery to toaster by an electrical current, students should define it as electrical energy. Students can determine that there is a current in the closed circuit by including an ammeter in it. They will learn to associate the current with the energy it carries by noting that they feel the heat from the toaster or light bulb only if the current is on.

Energy Transformations

Kinetic <—>Potential: Students should revisit examples of energy transformations and examine them quantitatively. Activities also could enable students to measure potential energies, velocities, and kinetic energies. Students should look at mechanical energy transformations as they occur in both directions, and they should consider what happens to the mechanical energy: Is it conserved, or is some of it lost?

Mechanical <—> Thermal: Several activities can illustrate conversions of mechanical energy to thermal energy. The only problem is that small increases in thermal energy may not increase noticeably the temperature of the system to which it is added. One method for investigating these types of conversions is to use thermistors or other sensitive sensors to detect small changes in temperature. An appropriate qualitative method would involve observing temperature increase when a thin piece of metal stops a rotating bicycle wheel.

Battery energy —>Thermal: To understand the transformation of battery energy to thermal energy, students could connect a battery to a wire resistor and feel the resistor get hot. A piece of wire can serve as a resistor, but unless it has a resistance comparable to the resistance of the battery, the thermal energy will heat the battery not the wire.

Conservation of Energy

Specific heats: The temperature increase of a system is not determined solely by the thermal energy added to it, but also by how much matter shares the energy and by certain intrinsic material properties. Students can explore this concept using a heat source that produces heat at a steady rate (a candle, hot plate, or an alcohol burner) and by revisiting investigations from Grades 6-8. These activities should be extended to be more quantitative and to illustrate the difference in the specific heats of water and some common metals. Students should analyze graphs of temperature versus time for different masses of water to see that, in each system, each gram of water requires the same amount of heat to increase its temperature one degree Celsius. By graphing the temperature change versus the time for different masses of metals (aluminum pellets, for example), students discover that the slope for the best straight line through the experimental points turns out to be larger than the slope for water. Thus, metal stores less energy per gram per degree of temperature increase than water does.

ELECTRICITY AND MAGNETISM

Moving Charge and Magnets

Ammeters and voltmeters: In Grades 6-8, students used the brightness of a bulb as a crude measure of electric current. In Grades 9-10, they should become familiar with the ammeter as a more quantitative measure of current. An ammeter and a bulb can be connected in a series, and students can observe that whenever the bulb is brighter, the ammeter reads a bigger current. Introducing the ampere as the unit for current, the teacher can discuss the calibration of an ammeter. Then, students should measure the current through several different circuits that include the same battery and note that the current is less for two bulbs in series than for one. A voltmeter can be added to the circuits so that students can measure the voltage of different batteries.

Faraday Induction: Students should experiment with energy conversion by connecting a wire coil to a sensitive ammeter or voltmeter, thrusting one end of a strong bar magnet through the coil, and watching the meter needle during the motion. Most electrical generators function by converting kinetic energy to electrical energy in this way.

Electric Circuits

Ohm's law: Activities should demonstrate Ohm's Law quantitatively. After constructing a circuit that includes a battery, a resistor, and an ammeter in series, and a voltmeter connected across the resistor, students should record the readings of both the current through the resistor and the voltage drop by adding another battery in the series or another resistor (but the voltmeter should be connected only across the original resistor). Students should repeat the activity to obtain different sets of values and then graph the current in amperes versus the voltage in volts. Students should realize that the reciprocal of the slope of the straight line through the points is the resistance, in ohms, of the original resistor.

Power law: Since the current is the charge per unit time through the resistor, and the voltage drop is energy delivered to the resistor per unit charge passing through it, the product of the two is energy per unit time and is called power. Students could calculate power by modifying the circuits used in the study of Ohm's law. But this time, they would immerse the resistor in a small amount of water in a plastic foam cup and insert a thermometer to measure the water temperature change. They should allow the current to flow long enough so that the water temperature increases by several degrees. Students record: (1) the current through the resistor, (2) the voltage drop across the resistor, (3) the time the current existed, and (4) the water temperature increase.

To obtain different current and voltage drops, students should repeat the above steps with circuits that have been altered to facilitate this activity. Students should be sure to use the same amount of water in the cup—at the same initial temperature—and to let the current remain on long enough to produce the same temperature increase as before. The product of current times voltage times time is the energy delivered to the water. Since the temperature increase is a measure of this energy, students should find the products to be approximately the same in both cases.

WAVES AND LIGHT

Mechanical Waves

Interference of water waves: The study of interference effects that are characteristic of waves should build on the study of water waves in Grades 6-8. Two periodic point disturbances, separated by a few centimeters, on the surface of water in a ripple tank reveal all the features of an interference pattern very clearly. Students should both observe and propose the origin of the nodal lines found in the patterns and the areas where crests and troughs propagate away from the wave sources. Students also should note the differences in the interference patterns produced when the distance between the point sources is changed and when the frequency of the source disturbances is altered to produce variations in the wavelengths of the water waves.

Sound

Interference: Activities with sound should enable students to draw the following conclusions: (1) sound is carried by a wave; (2) interference effects result when waves from two or more sources coexist in the same medium. Students also should associate these amplitude variations with the alternating paths of minimum and maximum disturbance (observed previously in the case of waves on a water surface).

Light: Characteristics and Models

Refraction: Refraction is the bending of a ray of light when it crosses the boundary between two transparent media. Students should observe refraction as a beam of light passes through various media, for example, air, water, mineral oil, transparent plastic, glass, blocks of gelatin, and salt or sugar solutions. Students should describe how the refracted ray bends at the interface and what happens when the angle of entry is changed. Additionally, teachers could help students see the reflected ray produced by the incident beam.

Total internal reflection: To explore total internal reflection, students could alter the angle of the ray of light incident to a glass/air, plastic/air, or gelatin/air interface so that the ray is not transmitted through the interface but is reflected back into the original medium. Total internal reflection is particularly dramatic when the reflection of the incident ray occurs throughout the original medium. Such an activity can serve as an introduction to fiber optics.

Interference: Interference patterns are observable only if the separation between the two sources is not significantly larger than the wavelength. Since the wavelength of visible light is small, interference patterns for light cannot be demonstrated as easily as interference patterns for water waves or sound. Students can look through two slits scratched in otherwise opaque material (painted glass, for example). Looking at a bright light source through such slits, students will see alternate bright and dark lines. Covering the light source with colored filters or gels produces different patterns. Students' observations should suggest that light is carried by a wave and that different wavelengths produce patterns that can be used for identification purposes.

Wave model of light: Students should describe how light carries energy from a source to a detector. The fact that light produces sharp shadows suggests that it may consist of particles that travel in straight lines. Once students see that light produces interference patterns, they could propose that it is a wave. However, all the other waves students encountered traveled through an elastic medium. What carries light waves? And what is waving? Students can explore these questions when they have completed further activities in Grades 11-12 on how light is generated, transmitted, and absorbed.

MATTER

Kinds and Characteristics

The electron: Much evidence indicates the existence of a low-mass, negatively–charged particle, that scientists call the electron, which is a constituent of all macroscopic matter. Students should learn that a beam of such particles is emitted by a metal filament when the filament is heated to a high temperature. Students also should encounter the Millikan oil drop experiment which showed that charge comes in fundamental units and measures the size (in charge units) of the smallest unit of charge.

Metals: Students should explore the following properties of metals: (1) electrons can be removed readily from metals; (2) a metal plates out on an electrode as positive copper ions accept electrons to become metallic copper during electrolysis of copper chloride solutions; (3) metals seem to have electrons weakly bound to massive positive ions; (4) electrons conduct electricity well; (5) metals conduct heat well, and they are shiny.

Students might consider whether it is the low-mass electrons or the high-mass ions that move when a current exists in a metal, and they should be challenged to propose a model that explains metallic properties. Teachers might initiate model making by suggesting that, in a metal, the weakly-bound electrons are free to move among a lattice of positive ions. However, it is premature to claim that this primitive model is consistent with the thermal and optical properties of metals, although it is clearly consistent with some of the electrical and chemical properties.

The atomic nucleus: Students should become familiar with the scattering experiments, originally conducted by Rutherford and his colleagues, that defined the early models of the atomic nucleus and that characterize the distribution of mass within an atom's volume.

Properties

Resistivity: Students already learned that metals conduct electricity much better than nonmetals. Students might examine this phenomenon quantitatively by measuring the resistances with many meters of fine copper wire. They should measure both the current through it and the voltage drop across it when connected to a battery. Students should discover that, for metals, resistance is independent of the current as long as the temperature does not change. They should explore how resistance varies with the wire length and is inversely proportional to the cross sectional area of the wire. Finally, students should determine that resistance is different for copper than for a different metal even though the length and cross sectional area are identical for both wire samples. The difference is related to the internal structure of the metal. Thus, students can conclude that a property called resistivity exists which can be measured for metals and used to characterize them. Students also should discover that nonmetals conduct even though their resistivities are several orders of magnitude higher than those of metals.

The size of atoms and nuclei: Numerous activities provide information on the size of atoms. Several activities should be used so that students are convinced by the convergence of different kinds of evidence. The teacher could coordinate this unit with appropriate units in chemistry. Of particular interest are electroplating experiments that measure the charge needed to plate out one mole of copper on the anode. Quantitative derivations from half-cell reactions can lead to determinations of the radius of a copper atom and suggest that a nucleus is smaller than the neutral atom by at least four orders of magnitude.

Radioactive half–life: The time it takes a particular radioactive nucleus to decay is different from one nucleus to another and cannot be predicted for any particular nucleus. However, students can engage in simulation activities of radioactive decay to produce graphs of the number of decays as a function of how long the nuclei lived. These can show readily that the curve is exponential and is characterized by the time it takes for half of the original nuclei number to decay. Geiger counters with radioactive samples and computer sensors can provide students with more quantitative data for studying radioactive half-life.

Change Processes

Ionization: Atoms release one or more electrons when heated to very high temperatures. Ionization may occur through this addition of thermal energy or by bombarding matter with either charged particles or electromagnetic radiation. It is important that students realize that all loosely bound electrons can be removed by adding just a few electron volts of energy to an atom.

Radioactivity: Some matter is radioactive. In other words, atoms of certain matter spontaneously emit radiation. Radiation exists in three types called alpha, beta, and gamma. Teachers should keep discussion of these particles simple: students only need to know that the particles are neutral and carry energy. No matter what the emitted particle, it carries an amount of energy at least 10^5 times larger than the ionization energy of an atom. Thus, radioactivity must be a nuclear phenomenon. Another piece of supporting evidence is that radioactive processes are unaffected by changes in the chemical state or physical conditions of the active atoms. Such changes alter the atoms' electron configurations but not their nuclear structures.

Models

The atomic model: Students should synthesize their activities on properties of atoms, the size of nuclei, electrostatic forces, binding energies, and fundamental particles to trace the evolution of atomic models from the simple one presented by Rutherford to the current Bohr model. Curricula could integrate investigative activities from the energy section and the study of light spectra so that students understand basic quantum numbers and energy levels. Students should speculate on the nucleus and the strong and weak forces. A brief introduction to elementary particles can be offered to familiarize students with the eightfold way and quarks.

P
H
Y
S
I
C
S

11-12

MOTION AND FORCE

Causes of Motion

Vector Addition: Students will need to add vectors in two dimensions. To begin activities, the teacher can discuss the properties of vectors and introduce simple vector addition. Students then should study motions that involve changes in displacement and velocity in two dimensions. Projectile motion and motion in a plane are appropriate areas for conducting activities.

Newton's second law: Students should build upon the qualitative understanding of this law developed in earlier grades. In particular, they should develop a quantitative understanding of Newton's second law as it applies to (1) objects moving under constant tensions that are not parallel to the direction of motion, (2) forces acting on an object in equilibrium, (3) forces acting on projectiles, (4) non-parallel forces acting on objects moving on a rough incline plane, and (5) velocity-dependent forces such as the resistance of a fluid to a falling object's motion.

Equilibrium

Stability of nuclei: Students previously found that some nuclei spontaneously decay by emitting alpha, beta, or gamma radiation. When this occurred with a short half–life, students determined that these were especially unstable nuclei. In Grades 11-12, students should consider experimental evidence on alpha decay, nucleon production, binding energies, and mass as a form of energy to derive a quantitative measure of stability. To discover trends in stability patterns that occur in the Periodic Table, students also should consider graphs of binding energy per nucleon versus number of nucleons.

Nuclear forces: Students should be curious about what causes protons and neutrons to stick together in a nucleus. The answer is, the strong nuclear force. Students should be aware of the assumption that the nuclear force between any two nucleons is small except when their separation is about 10^{-15} m. At this distance the force is very strong and attractive while at larger separations, the force rapidly drops to zero. Students should compare the strong nuclear force with the weak nuclear force as well as with the other fundamental forces.

Momentum: A Conserved Quantity

Conservation of momentum (quantitative in two dimensions): Students should review Newton's third law and note that it guarantees the conservation of momentum in any two-body collision. Students should consider the following kinds of collisions: (1) One–dimensional, totally inelastic; (2) One–dimensional, elastic; (3) Two–dimensional, inelastic (Car collisions at an intersection are interesting examples); (4) Two–dimensional, elastic.

ENERGY

Kinds of Energy

Radiant energy: Students should examine the electromagnetic spectrum and consider both the wave and photon models. Introductory activities can focus on radio waves and how they are generated, transmitted, and received. Since the source of this energy is an oscillating electrical circuit, students should become convinced that electrical and magnetic effects are strongly involved in the transmission of radio waves. The teacher should point out that the measured velocity of these waves is 3×10^8 m/s, which is also the velocity of light. Students need to develop the relationships between velocity, frequency, and wavelength. Additional activities should incorporate visible light as a wave source, and final activities should address gamma rays emitted by radioactive nuclei. Instruments such as Geiger counters can detect emitted photons, and their speed and energy can be derived. Students should consider the theory developed by Planck and Einstein which proposed that radio waves, visible light, and gamma rays—as well as other forms of electromagnetic radiation—are all examples of energy carried by photons. Collectively, though, photons exhibit wave characteristics and have interactions thought of as electromagnetic in nature.

Energy Transformations

Fission: When a heavy nucleus, such as uranium, decays naturally, it emits a low–mass particle (an alpha or beta ray) and changes into a different, heavy nucleus. Fission occurs when a heavy nucleus splits roughly in half. That such an event is possible energetically follows from the fact that nuclei in the middle of the Periodic Table are the most stable. Students might explore the experimental material available on fission and chain reactions. The class also could discuss the advantages and disadvantages of nuclear reactors as energy sources.

Fusion: Previously having considered stability trends in the Periodic Table and fission reactions, students now can consider fusion reactions as low-mass nuclei combining to release energy. Students can research current efforts to generate energy from fusion reactions and speculate about the future of fusion devices as energy sources for the world.

Conservation of Energy

Quantized energy levels: A major puzzle to 19th century physicists was the line spectra emitted by the atoms of gases. Neil Bohr provided a theoretical explanation when he proposed the nuclear model of the atom and the idea that atomic energies are quantized. The 11-12 grade curricula should coordinate physics and chemistry activities so that students can determine the energy levels for atoms using visible line spectra emitted by heating gases.

Conservation of mechanical energy: In earlier grades, students observed transitions from kinetic energy to potential and back again. Students should revisit these activities and carefully measure potential and kinetic energies to confirm the existence of a conservation law. Rotational and translational motion, friction, and the loss of mechanical energy should be considered.

Mechanical equivalent of heat: Activities should explore the transition from mechanical energy to thermal energy. Historically, however, mechanical energy units (joules) and thermal units (calories) were invented separately. Therefore, early experiments were measures of the conversion factor. They also can be viewed as evidence for the conservation of energy if it turns out that the conversion factor determined in this way never varies. Students can return to earlier activities and determine the conversion factor in a more quantitative manner. From their attempts to transform electrical energy to thermal energy, students should find that one calorie is equivalent to 4.186 joules.

First law of thermodynamics: The first law of thermodynamics is the conservation of energy phrased in terms of work, heat, and internal energy. If students place a flame beneath a container of gas, heat is conducted through the walls to the gas. The energy of the gas increases. This could be termed internal energy. Students can measure this increase with a thermometer and consider it to be the same as what previously was called thermal energy. Students then can remove the flame and insulate the container so that no heat can enter or exit. One wall of the container should be moveable: students push it in, thereby compressing the gas. Work was done on the gas, and its internal energy increased an equivalent amount. Students again could measure this increase with a thermometer. Combining the above results enables students to conduct the calculations that yield the first law.

Second Law of Thermodynamics

Heat engines and refrigerators: Students can understand the second law of thermodynamics by considering how heat engines and refrigerators work. Students can imagine a heat engine that operates in a closed cycle so that series of processes can be repeated. Each cycle should involve adding net heat and extracting an equal amount of work. If heat enters the system from a source at a high temperature, a fraction of this heat must be rejected to the surroundings, and therefore all heat input cannot convert to useful work. The rejected heat still exists as thermal energy, but, because of the degradation in temperature of the matter storing this energy, the energy is less available for doing work in the future. Students should recognize this one-way nature of energy degradation is a hallmark of the second law.

Students can consider a refrigerator as a heat engine running backward. Doing work on the system can remove heat from cold matter and deposit it in the surroundings that are at a higher temperature. In this context, the second law simply asserts that without doing the work, one could not pump heat up a thermal gradient. Students should see that heat always flows from hotter to colder places—a simple and intuitive form of the second law.

ELECTRICITY AND MAGNETISM

Static Charge

Coulomb's law: Students already have explored the characteristics of electrostatic forces, but they should build these basic ideas into a quantitative theory. To establish Coulomb's law, the class can use simple equipment or instructive video materials commercially available.

Moving Charge and Magnets

Magnetic forces on moving charges: Cathode ray tube activities can demonstrate that magnets exert forces on moving charged particles. Students could tune the cathode ray tube so that a spot of light is near the center of the tube face. They then can bring one end of a strong bar magnet near the tube about halfway between the ends. The beam of electrons will be deflected, and the spot of light will move.

Magnetic forces on currents: Activities should demonstrate that magnets exert forces on current–carrying wires. A flexible wire carrying a current of about an ampere will move when placed in the field of a strong horseshoe magnet. Forces like these cause the solenoid valves in washing machines and other appliances to operate and electric motors to turn.

Electric Circuits

Parallel circuits: In earlier grades, students studied series circuits. The teacher could introduce parallel circuits by defining that two or more branches are in parallel when they provide alternate paths for charge to move from point a in a circuit to point b. Students then should experiment with different arrangements of resistors in series and parallel. They should measure currents and voltage drops within circuits to uncover patterns for resistance that exist in series and parallel circuits.

Fields

Static and magnetic fields: Whenever one object exerts a position-dependent force on other objects, scientists say that the first object creates a force field around it, and the other objects feel this force whenever they enter the region where this field exists. For example, the Earth produces a gravitational force field around it which differs from point to point in both magnitude and direction. Once this field is compiled, scientists can determine the force on any mass placed at any point near the Earth. Students may not be able to measure different effects of the gravitational force field within the classroom, but they can explore fields by experimenting with electric charges. Students also can describe magnetic interactions in terms of fields.

PHYSICS

11-12

WAVES AND LIGHT

Sound

The Doppler effect: When a source of sound or a sound detector moves relative to the medium through which the sound travels, the detector measures a different frequency than would be observed if there were no motion. The frequency is higher if the source and detector approach each other, and lower if they separate. Students undoubtedly have heard this effect when trains or ambulances pass them while sounding a whistle or siren. Students should replicate the effect with available equipment. Students also should note that a similar effect exists for light. This is particularly important to astrophysicists because the effect allows them to measure the velocities with which distant galaxies recede from our own.

Light: Characteristics and Models

Light as an electromagnetic wave: When students studied electromagnetic fields, they learned that whenever electric charges oscillate, they create time-varying electric and magnetic fields that propagate away from the source as a wave.

The photon model of light: While studying energy, students learned that physicists sometimes treat light as a stream of photons, where each photon carries a discrete amount of energy and momentum. In previous activities on quantized energy levels, students observed line spectra emitted by glowing gases. In order to build on the photon model of light, students now should make the connection between the photon's energy and the wavelength measured by passing photons through a spectrometer. This connection enables them to gain insight into the correspondence between energy, Planck's constant, the velocity of light in vacuum, and the photon's wavelength.

Merging the models: Students now should be as confused as scientists were when these different models were developed. One resolution suggests that light is a stream of particles but recognizes that a single photon's behavior is governed by statistical laws (the laws of quantum mechanics), not by laws that link cause and effect in terms of a specified process. For example, when an individual photon of a certain energy passes through a slit with a width comparable to the wavelength associated with a photon of this energy, it can emerge with different amounts of transverse momentum. Thus, this photon can be detected at different points on a viewing screen. The statistical probabilities for arrival at a particular point are such that, collectively, many photons form interference patterns characteristic of a wave. This concept is a difficult one for beginners and even for experienced scientists. The teacher should stimulate students simply to think and ask questions about the statistical nature of particle behavior. Perhaps students can recall examples from earlier studies such as radioactive decay. No one can predict when any particular nucleus will decay, but one can predict the fraction that will decay in a particular time interval.

BIBLIOGRAPHY

Supporting and Documentary Materials

Aldridge, Bill G. 1989. "Essential Changes in Secondary Science: Scope, Sequence and Coordination." **NSTA Reports!** (January): 1, 4– 5.

Founding article for the SS&C project. Identifies the inadequacies of American secondary school science education and calls for reform. Proposes 1) the study of science every year for six years by all students; 2) coordination of the four disciplines so that students see the interrelationships and applications of important concepts; and 3) spaced learning— revisiting concepts on a periodic basis, giving students the opportunity for increasingly rich and deep levels of comprehension. In the proposed sequenced approach, science education would move from the descriptive and phenomenological, to the empirical and semi–quantitative, and finally to the theoretical and abstract.

Aldridge, Bill G. 1992. "Scope, Sequence and Coordination: A New Synthesis for Improving Science Education." **Journal of Science Education and Technology** (Spring).

Provides a brief history of the Project on Scope, Sequence, and Coordination of Secondary School Science. The article details both the actual project and its assessment component, a prototype compact disc interactive (CD– I) .

Anderson, Charles W. 1987. "Strategic Teaching in Science." In **Strategic Teaching and Learning: Cognitive Instruction in the Content Areas.** Eds. Jones, Beau, Fly et al. Alexandria, VA: Association for Supervision and Curriculum Development.

Contrasts teaching science as memorizing facts, rules, and definitions and teaching science as understanding how and why events occur as they do in the natural world. Anderson does not reject the explicit teaching of science concepts, but suggests that science concepts be put in the context of meaningful problems. Students then can ask their own questions, realize their own misconceptions, and gradually reconstruct their understanding of how the world works. Additionally, Anderson believes each important concept should be included in different tasks and explicitly identified.

Arons, Arnold B. 1983. "Achieving Wider Scientific Literacy." **Daedulus** 112 (2): 91– 122.

Enumerates the attributes that scientifically literate students would possess. Arons says it is essential to cover less content and to slow the learning pace so that students can "follow and absorb" a few key science concepts. Supports students' consideration of the questions, "How do we know...? Why do we believe...? What is the evidence for...?"

Bruner, Jerome S. 1960. **The Process of Education.** Cambridge, MA: Harvard University Press.

Built on the premise that any student at any age can learn science ideas and themes in some form. States that in early grades, students can and should intuitively grasp and use basic ideas. In later grades, they can encounter the same ideas in progressively complex forms and thus, continually broaden and deepen their knowledge. Bruner discusses a spiral curriculum and supports the development of curricula that revisit basic concepts and build upon them.

Brunkhorst, Bonnie J. 1991. "Every Science, Every Year." **Educational Leadership** (October): 36– 38.

Describes implementation of SS&C in a California high school, featuring the experiences of the curriculum designers and teachers. Identifies methods that were and could be used for achieving coordination.

Dellarosa, Denise and Lyle E. Bourne, Jr. 1985. "Surface Form and the Spacing Effect." **Memory and Cognition** 13(6): 529– 537.

Critiques and elaborates previous psychological research on the spacing effect. Presents own results which indicate that recall of any item is improved when it receives full processing (achieved by spacing and by varying surface features).

Dempster, Frank N. 1988. "The Spacing Effect: A Case Study in the Failure to Apply the Results of Psychological Research." **American Psychologist** 43(8): 627– 634.

Presents psychological research that, for a given amount of study time, spaced presentations are substantially more effective in increasing learning than intense, singular sessions. Dempster argues for using this knowledge of "spaced learning" for instructional purposes in the development of curricular frameworks.

Kahle, Jean Butler. 1990. "Why Girls Don't Know." **What Research Says to the Science Teacher** 6: 55– 67.

Using a variety of research studies, Kahle examines the scarceness of women in science and proposes some methods for encouraging and retaining them in science programs. She identifies the unique difficulties that women encounter on the elementary, secondary, and college/university levels. The article goes beyond simply discussing the problem of recruiting and retaining women students in science to consider equitable science education for both women and men.

Manning, M. Lee and Robert Lucking. 1991. "The What, Why, and How of Cooperative Learning." **The Clearinghouse** 64 (3): 152– 3.

Takes up the issue of the instructional needs of the heterogeneous student body. Manning and Lucking discuss cooperative learning as an alternative to traditional grouping plans and instructional systems. They define cooperative learning as "techniques in which students work in heterogeneous groups of four or six members and earn recognition, rewards, and sometimes grades based on the academic performance of their groups." They discuss evidence indicating that cooperative learning can contribute positively to students' academic achievement, social skills, and self–esteem.

Minstrell, James A. 1989. "Teaching Science for Understanding." **Toward the Thinking Curriculum: Current Cognitive Research, 1989 ASCD Yearbook.** Eds. Lauren Resnick and Leopold E. Klopfer. Alexandria, VA: Association for Supervision and Curriculum Development.

Presents a sample physics lesson using conceptual change principles. Discusses more generally what the conceptual change principles suggest for helping students "restructure" knowledge, for the design of instruction, and for fostering an appropriate classroom environment for learning.

Orrill, Robert, ed. 1990. **Academic Preparation in Science: Teaching for Transition from High School to College.** New York: College Entrance Examination Board.

Compiles recent findings on the cognitive processes involved in learning science and discusses what the findings suggest for student achievement of desired learning outcomes. Based on the findings, the College Board calls for science instruction that uses a "structured inquiry" approach, integrates content and processes in science curricula, and involves teachers as "facilitators." In a very accessible format, this book describes why restructuring should be undertaken and points educators in the right direction for undertaking it.

Piaget, Jean. 1973. **To Understand is to Invent: The Future of Education.** New York: Grossman Publishers.

Presents the argument that students, to develop cognitively, must have direct experience with phenomena and construct their own explanations. Piaget's "constructivist" approach to concept development, outlined here, subsequently received significant support from research in the cognitive sciences.

Rakow, Steven J. 1986. **Teaching Science as Inquiry.** Bloomington, IN: Phi Delta Kappa Educational Foundation.

Suggests that students need to acquire science process skills. These skills model the investigative processes that scientists use in developing scientific concepts. Rakow asserts that science programs must focus on facilitating the development of students' science process skills through hands–on investigations and experiments.

Resnick, Lauren B. 1987. **Education and Learning to Think.** Report of the Committee on Mathematics, Science and Technology Education, National Research Council. Washington, DC: National Academy Press.

Defines higher order thinking skills and suggests how schools can facilitate their development in students. The monograph also considers the nature of thinking and learning, general reasoning, and the improvement of intelligence.

"Revamping Science Education is No Piece of Cake." 1991. **Chemecology** 20(3): 2–3.

Briefly outlines the philosophy of the SS&C project and describes its initial implementation in four Houston, Texas schools and 110 California schools.

Reynolds, James H. and Robert Glaser. 1964. "Effects of Repetition and Spaced Review Upon Retention of a Complex Learning Task." **Journal of Educational Psychology** 55(5): 297–308.

Presents the initial documentation of the spacing effect. Reynolds and Glaser's results indicated that spaced review significantly improves retention of reviewed material.

Roth, Kathleen J. 1989. "Science Education: It's Not Enough to 'Do' or 'Relate.'" **The American Educator** (Winter).

Considers and evaluates three perspectives in science teaching— the inquiry perspective, the Science–Technology–Society (STS) perspective, and the conceptual change perspective. Finds that the conceptual change perspective offers the most useful framework for science instruction because it encourages students to use processes to develop better understandings of natural phenomena.

Rutherford, F. James and Andrew Ahlgren. 1990. **Science For All Americans: A Project 2061 Report on Literacy Goals in Science, Mathematics, and Technology.** New York: Oxford University Press.

Presents a set of recommendations for reforming the content and character of American education in science, mathematics, and technology. Defines the knowledge, skills, and attitudes all students should acquire from K–12 education.

Advocates reducing content, emphasizing connections among traditional subject matter categories, and developing students' "ideas and thinking skills" instead of specialized vocabulary.

Sachse, Thomas P. 1991. "Commonalities in the NSF Scope, Sequence and Coordination Projects: How the Six Centers Comprise a National Reform." Sacramento, CA: California Department of Education.

Reviews the early progress of the six SS&C implementation sites with attention given to shared features and similar approaches to handling the challenges posed by the restructuring process. Specifically addresses curriculum, instruction, involvement, quality control, and systematic reform.

Scope, Sequence and Coordination of Secondary School Science. 1992. **Volume II: Relevant Research.** Washington, DC: National Science Teachers Association.

Companion volume to The Content Core. Makes available to the science educator a select group of research papers on how secondary school students learn science best. Its papers address current learning research issues such as spacing content, cooperative learning, equitable classrooms, cognitive skill development, and conceptual change. Chapter introductions indicate connections between the research evidence and the tenets of SS&C.

Slavin, Robert E. 1990. "Achievement Effects of Ability Grouping in Secondary Schools: A Best–Evidence Synthesis." Madison, WI: University of Wisconsin–Madison, National Center on Effective Secondary Schools.

Finds that between–class ability grouping does not improve learning and recommends that such grouping be reduced.

Songer, Nancy Butler and Marcia C. Linn. 1991. "How Do Students' Views of Science Influence Knowledge Integration?" **Journal of Research in Science Teaching** 8 (9): 761-784.

Examines middle school students and finds that students with a "dynamic" view of science acquire a more integrated understanding of science principles than those who possess "static" views. Songer and Linn interpret their results to mean that if science courses avoided presenting pieces of scientific knowledge in isolation and instead emphasized knowledge integration and consideration of the nature of science, then students would develop a more cohesive, productive understanding of science.

von Glasersfeld, Ernst. forthcoming. "Questions and Answers About Radical Constructivism." **The Practice of Constructivism in Science and Mathematics.** Washington, DC: AAAS Press.

Von Glasersfeld responds to common questions about constructivism and discusses the educational implications of adopting a radical constructivist view. A key observation he makes concerning science education is that hypotheses, natural laws, and other scientific theories should not be taught as "truth" but rather as the "most viable" explanations for phenomena.

Yager, Robert E. 1991. "The Constructivist Learning Model: A Basis for Real Reform in Education." Iowa City, IA: University of Iowa, Science Education Center.

Characterizes the constructivist learning model and its potential for changing teaching methods. This model operates on the premise that learning outcomes are the interactive result of the information students encounter and how they process it based on preconceived notions and existing personal knowledge. Proposes how science teachers could move to more constructive approaches.

National Reports

Action Council on Minority Education. 1990. **Education That Works: An Action Plan for the Education of Minorities.** Cambridge, MA: Quality Education for Minorities Project, Massachusetts Institute of Technology.

Comprehensively discusses the educational needs of minorities and sets goals for meeting those needs through individual and national allocations of energy, resources, and commitment.

Blumberg, Fran, M. Epstein, W. MacDonald, and Ina V.S. Mullis. 1986. **A Pilot Study of Higher Order Thinking Skills Assessment Techniques in Science and Mathematics: Final Report, Part 1.** National Assessment of Educational Progress. Princeton, NJ: Educational Testing Service.

Develops a conceptual framework of higher order thinking skills in science and mathematics which was used to construct prototype exercises, including hands–on activities in which students were asked to solve problems, conduct investigations, or respond to questions using materials and equipment. Presents problems and results of the pilot tests.

Blumberg, Fran, M. Epstein, W. MacDonald, and Ina V.S. Mullis. 1986. **A Pilot Study of Higher Order Thinking Skills Assessment Techniques in Science and Mathematics: Final Report, Part II, Pilot Tested Tasks.** National Assessment of Educational Progress. Princeton, NJ: Educational Testing Service.

Presents the instruments and samples of student responses to problems used in the study to develop ways to assess higher order thinking skills at grades three, seven, and eleven in both science and mathematics. Pilot–tested tasks included group tasks, station activities, and individual investigations.

Carnegie Council on Adolescent Development. 1989. **Turning Points: Preparing American Youth for the 21st Century.** Washington, DC: The Council.

Summarizes the condition of adolescents in the 1980s and the need for changing educational structures and practices. The central section, dealing with "Transforming the Education of Young Adolescents," proposes creating learning communities in middle schools by teaching a core academic program.

National Research Council. 1990. **Fulfilling the Promise: Biology Education in the Nation's Schools.** Washington, DC: National Academy Press.

With a focus on biology, this report addresses many issues relevant to restructuring science education: the history of science education, the need for change, obstacles to change, how to encourage and support good science teaching, and the need for leadership. Includes especially notable discussions on reducing content and the importance of laboratory activities.

National Science Board Commission on PreCollege Education in Mathematics, Science and Technology. 1983. **Educating Americans for the 21st Century: A Plan of Action for Improving Mathematics, Science and Technology Education for All American Elementary and Secondary Students so that their Achievement is the Best in the World by 1995.** Washington, DC: National Science Foundation.

Proposes student outcomes for science, mathematics, and technology education at the elementary, middle, and senior high school levels. Presents recommendations for the continuing education of science and mathematics teachers and for changes in preservice teacher education.

Oakes, Jeannie. 1990. **Multiplying Inequalities: The Effects of Race, Social Class, and Tracking on Opportunities to Learn Mathematics and Science.** Santa Monica, CA: Rand Corporation.

Finds that race, social class, and tracking limit the opportunities of students to achieve and participate in science and mathematics classes. Oakes specifically considers how these characteristics relate or correspond to differences in the quantity or quality of science and mathematics curricula.

The Task Force on Women, Minorities, and the Handicapped in Science and Technology. 1989. **Changing America: The New Face of Science and Engineering.** Washington, DC: National Science Foundation.

Offers specific actions for those key groups of decision makers who set educational policy. Actions focus on keeping underrepresented groups in the science and engineering pipeline. Recommends actions for educators, including: "Make science hands–on. Ensure that all students do science as well as read about it."

ACKNOWLEDGMENTS

During the development of **The Content Core**, many individuals contributed to its design and were involved in the review of different drafts. We would like to acknowledge, with gratitude, their participation. Some of these teachers, scientists, educators, and supervisors may not recognize their contributions, but those who have followed the project understand that each led to a more useful document.

David M. Andrews, Wilmington, NC
Arnold Arons, Seattle, WA
Grace Beam, Houston, TX
Glen Bennett, College Park, MD
Diane Beverley, Claremont, CA
Ronald J. Bonnstetter, Lincoln, NE
Bonnie J. Brunkhorst, San Bernadino, CA
Herbert Brunkhorst, San Bernadino, CA
Donald Cammiso, Fairfax, VA
John R. Carpenter, Columbia, SC
Arthur Christensen, Mahwah, NJ
Robert Christman, Bellingham, WA
Charles R. Coble, Greenville, NC
Linda W. Crow, Houston, TX
Jane Crowder, Isaquah, WA
Rebecca R. Dewey, Falls Church, VA
Arthur Eisenkraft, Bedford, NY
Fred Finley, Minneapolis, MN
Barbara Foots, Houston, TX
Dorothy L. Gabel, Bloomington, IN
Marjorie Gardner, Berkeley, CA
J. David Gavenda, Austin, TX
Anne George, Baltimore, MD
Manuel Gomez, San Juan, PR
Charles G. Groat, Baton Rouge, LA
Mary Gromko, Denver, CO
Judith Grumbacher, Alexandria, VA
Joan W. Hall, Cincinnati, OH
Garrett Hardin, Santa Barbara, CA
William K. Hartmann, Tucson, AZ
Gayle Hartmann, Tucson, AZ
Henry Heikkinen, Greeley, CO

Robert L. Heller, Duluth, MN
Darrel Hoff, Cambridge, MA
Frank W. Ireton, Washington, DC
Jane Butler Kahle, Oxford, OH
Shirley Kelly, Belle, WV
LeRoy Lee, Madison, WI
Arthur Livermore, College Park, MD
J. David Lockard, College Park, MD
Shirley Malcom, Washington, DC
Robert Manka, Washington, DC
Carolee Matsumoto, Newton, MA
Floyd E. Mattheis, Greenville, NC
James Minstrell, Mercer, WA
Charlles Misner, College Park, MD
Penny Moore, Piedmont, CA
Richard Nicholson, Washington, DC
Michael J. Padilla, Athens, GA
Robert Parry, Salt Lake City, UT
John E. Penick, Iowa City, IA
Roger Pense, Benicia, CA
Judy Philippides, College Park, MD
Jerry Pine, Pasadena, CA
Patricia A. Rourke, Alexandria, VA
Thomas P. Sachse, Sacramento, CA
Gary Sampson, Wauwatosa, WA
Jane Sisk, Murray, KY
Patricia Smith, Colorado Springs, CO
Teny Topalian, Long Beach, CA
Emma L. Walton, Anchorage, AL
Gerald Wheeler, Bozeman, MT
Ann L. Wild, Washington, DC
Robert E. Yager, Iowa City, IA

INDEX